THEORY OF MACHINES

Volume-I

(KINEMATICS OF MACHINES)

by

Prof. (Dr.) R.K. Singal
Educationist and Technologist

S.K. KATARIA & SONS®
Publisher of Engineering Books
4885/109, Prakash Mahal, Dr. Subhash Bhargav Lane,
Daryaganj, New Delhi-110002 (India)
Phone: +91-11-23243489, 43551243; Mobile: +91-9871775858
e-mail: katariabook@gmail.com, katariabooks@yahoo.com
Website: www.skkatariaandsons.com
(e-books, subscriptions and bulk buying also available on our website)

Published by :
S.K. KATARIA & SONS®
4885/109, Prakash Mahal, Dr. Subhash Bhargav Lane,
Daryaganj, New Delhi-110002 (India)
Phone: +91-11-23243489, 43551243; Mobile: +91-9871775858
e-mail: katariabook@gmail.com, katariabooks@yahoo.com

First Edition: August, 2012, Reprint: 2022

Price: ₹ 225/-

Laser Type Setting
PLUS COMPUTERS, Shahdara, Delhi-93

Printed at
New A.S. Printing Press, Delhi-110006

Preface...

Theory of Machines is a very important subject for the students of Mechanical Engineering. It deals with engineering theory, analysis, design and practice and kinematics and dynamics of machines. U.P. Technical University has divided the subject into two parts

1. Theory of Machines-I (Kinematics of Machines)
2. Theory of Machines-II (Dynamics of Machines)

The present book has been prepared as per syllabus EME-502 Theory of Machines-I (Kinematics of Machines) for the mechanical engineering students of Gautam Budh Technical University, Lucknow; Maha Maya Technical University, Noida, and Uttrakhand Technical University, Dehradun. Of course, the book will be useful for the students of other technical universities. The book has been written in simple language amply supported with neat drawings and appropriate solved numerical examples. Theory and Numerical Questions have been added for the practice of the students.

The author is grateful to the publisher for bringing out this book in a very short time. Although the book has been prepared keeping in mind difficulties and requirements of the students, any constructive criticism and suggestions will be welcome with gratitude.

Author

Dedicated to...

MAN AND HIS MACHINES

Man lived a life of miseries and drudgery,
Enslaved to work hard by others trickery.

"Had no time to stand beneath the bows,
And gaze as long as sheep and cows*".

Machines were invented for Industrial Revolution,
And provide leisure and life of spiritual evolution.

Work of thousands was done by machine powers,
Tasks were finished in seconds meant for hours.

Machines performed work with killing speeds,
But man rushed for gold with cunning greeds.

His machines are consuming resources with rapid pace,
Not even sparing for him breathing space.

Man has been enslaved by his own machines,
No time for leisure and pleasure in busy routines.

No time to sit and do meditation,
To study the mystic designs of the creation.

R.K.Singal

* Willam Wordswoth

Contents...

UNIT–I

1. SIMPLE MECHANISMS

2. VELOCITY ANALYSIS IN MECHANISMS

CHAPTER 1

SIMPLE MECHANISMS

1.1. INTRODUCTION

Man works to avoid work and has created machines and other mechanical systems to do his work. He has designed the machines with suitable mechanical advantage and velocity ratio to multiply the efforts of his muscle power and control the speed of the work output. Before the Industrial revolution of the 18th century, most energy used relied on human and animal muscles. The other sources of energy available in nature were kinetic energy of wind, potential energy of water and chemical energy in firewood and fossil fuels like coal, petrol and diesel. He designed engines and turbines to convert the natural sources of energy into useable form of electricity and mechanical energy of shaft notation and piston work. Modern high capacity, high speed machines work with the new forms of usable energy developed.

There are two aspects of the study of mechanical systems or machines, *i.e.*, **design** and **analysis**. The design or synthesis is the process of prescribing the sizes, shapes, material composition and arrangements of parts so that the resulting machine will perform the prescribed task. The designer needs large collection of mathematical and scientific tools and other reliable information. Analysis is the technique to critically evaluate and rate the designs already conceived or existing with the help of mathematical and scientific tools.

The **theory of machines** is an applied science that is used to understand the relationships between the geometry and motions of parts of a machine and the forces that produce these motions. Engineering mechanics is the branch of scientific analysis that deals with motions, time and forces. Engineering mechanics is also called stractural mechanics or rigid-body mechanics and the body is idealized as a continuum (continuous distortion of matter), that undergoes theoretically no deformation whatsoever. In actual practice, no solid body is perfectly rigid and everybody undergoes some deformation (change of shape and size) under the action of force. However, the deformation may be very small and may not be considered.

In the study of dynamic behavior of a rigid body the latter may be at rest and in equilibrium or it way be in motion. Statics deals with the behavior of bodies at rest and dynamics deals with the bodies in motion under the action of forces. Dynamics is also made up of two major disciplines as recognized by Euler in 1775.

"The investigation of the motion of a rigid body may be conveniently separated into two part, the one geometrical, the other mechanical. In the first part, the transference of the body from a given position to any other position must be investigated without respect to the causes of the motion, and must be represented by analytical formulae, which will define the position of each point of the body. This investigation will therefore be referable solely to geometry, or rather to stereotomy.

It is clear that by the separation of this part of question from the other, which belongs properly to mechanics, the determination of the motion from dynamic principles will be made much easier than if the two parts were undertaken conjointly".

These two aspects of dynamics are called **kinematics** and **kinetics**, Kinematics is the study of position, displacement, rotation, speed, velocity and acceleration. This is the initial problem in the design of a mechanical system and is investigated without respect to causes of motion and analytical formulae refer solely to geometry. Kinetics deals with motion and forces producing it. The determination of motion from dynamic principles will be made much easier if undertaken separately.

The classification of mechanics based on above discussion into major disciplines is shown in Fig. 1.1.

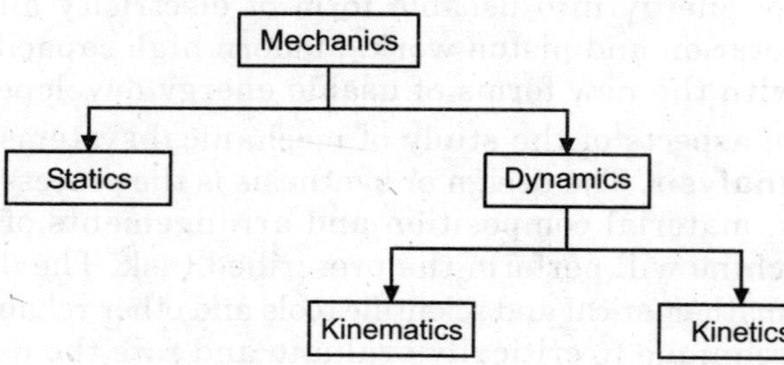

Fig. 1.1. Classification of mechanics.

1.2. CLASSIFICATION OF MECHANICAL SYSTEMS

The design and analysis of mechanical systems are based on the laws and principles of mechanics. There is a diret analogy between the various types of mechanical system with the three branches of mechanics, *i.e.*, statics, kinematics and kinetics. Therefore, these systems have been classified by taking analogy of classification of mechanics. The classification of mechanical systems is imposed on Fig. 1.1 and redrawn as Fig. 1.2.

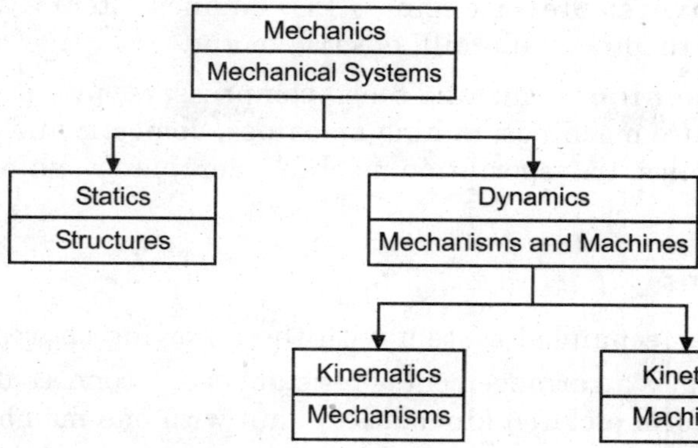

Fig. 1.2. Classification of mechanical systems.

The mechanical systems have been classified into the following equipment.

1. Structures
2. Mechanisms
3. Machines

1.2.1. Structures

A structure is a mechanical system or equipment with the following characteristics.

1. A structure is a combination of rigid (resistant) bodies connected by joints.
2. The structures is intended to be rigid and all joints are rigid.
3. The main purpose of a structures is to carry loads or forces and transmit or transfer the same to the foundation or supports.
4. A structures does not do any work or transform motion.
5. It has no internal mobility and there is no relative motion between its various members.
6. A structures can be shifted or moved from one place to another as a total assembly.
7. The main examples of structures are trusses, frames, columns, struts, beams, fixed members of mechanisms, stator parts of a machines.
8. A structures is designed on the basis of laws of statics.

1.2.2. Machines

A machine is a mechanical system with the following characteristics.

1. A machine is an arrangement of parts connected by movable joints.
2. There is interval mobility and relative motion between the members which is utilized to transforms motion and produce work.
3. A machine is intended to do work and transform motion.
4. It is a device for applying power or changing its direction.
5. The machine is described by the terms force, torque, work, power, etc.
6. A machine is deigned by the principles of kinetics.
7. The main examples of machines are given below.

***(a)* Prime movers:** Steam engines, I.C. engines, steam turbine, gas turbine, hydraulic turbine, windmill, electric motor.

***(b)* Driven machines:** Machine tools, compressors, pumps, industrial machines, agriculture machines, mining machines, domestic and commercial machines, automobiles, transportation machine, aeroplanes, ships, railways, etc.

1.2.3. Mechanisms

A mechanism is a mechanical system with the following characteristics.

1. A mechanism is an arrangement of resistant bodies connected by movable joints and form a closed kinematics chain with one member fixed.
2. There is internal mobility and relative motion between the members.
3. The purpose of a mechanisms is to transform motion by using the relative internal motions.
4. A mechanism may be a part of a machine.
5. It is designed by the principles of kinematics.
6. A mechanism may transmit power or force but its main purpose is to achieve desired motion.
7. The slider crank mechanism used in all reciprocating engines, pumps, compressors, etc. is an important example of a mechanism.

The present book discusses the kinematics of mechanisms whereas kinetics of machines is discussed in next volume. The statics of structures' is discussed in engineering mechanics.

1.3. ARRANGEMENT OF A SIMPLE MECHANISM

Reuleaux defines a machine as a **"combination of resistant bodies so arranged that by their means the mechanical forces of nature can be compelled to do work accompanied by certain determinate motions".**

He has also defined a mechanism as an **"assemblage of resistant bodies, connected by movable joints, to form a closed kinematic chain with one link fixed and having the purpose transforming motion".**

The above definition can be used to describe the arrangement or construction of a mechanism as follows.

1. The component of a mechanism or a machine is called a **link** which is assumed to be completely rigid.
2. The links of a mechanism are connected together is a suitable manner in order to transmit motion from the driver, or input link to the follower or output link.
3. The joints and connections between the links, are called **kinematic pairs**.
4. **Kinematic chain** is formed by joining some links by moveable joints. A closed loop is formed by joining every link of a chain to atleast two other links.

5. A **closed kinematics chain** is formed with one link fixed which is used to measure the relative motions of other points on the linkage. The fixed link is called the frame or **base link**.
6. A kinematic chain is called a **mechanism** when the frame link is specified.
7. Proper relative motions between the links are produced by proper selection of the number of links and the types of joints to connect them.
8. **Pair variables** depend upon the degrees of freedom of the joint selected. The relative motion between kinematic pairs is a function of pair variables.
9. If there is a surface contact between the pair elements the kinematic pair is called **lower pair**.
10. The **higher pairs** have line or point contact between the pair elements.
11. A mechanism may refer to a wide variety of devices including both higher pairs and lower pairs. A mechanism having only lower pairs is also called a **linkage**.

1.3.1. Types of Links

A component of a mechanism or a machine part is designated as a **link**. A links has to perform the following functions.

1. The link should be rigid so that there is no relative motion or change of distance between any two selected points on the same link.
2. It should hold constant spatial relationships between elements of a kinematic pair. It should have relative motion.
3. The link should be a resistant body and capable of transmitting the required forces with negligible deformation.

The following types of links can be used to transmit motion between the driver and the follower.

1. **Rigid link** can transmit motion without any deformation. Examples: connecting rod, crank of an I.C. engine.
2. **Composite rigid link** may consist of several parts rigidly fastened together without any relative movement between any two points on composite link. Examples: connecting rod with big and small bearings; crank, crankshaft and flywheel; cylinder, engineer frame and main bearings; piston, piston rod and crosshead.
3. **Flexible link** can partly deform in a manner that does not affect the transmission of motion: Examples: belts, ropes, chains and wires are flexible links which are rigid when transmitting motion without deformation while under tensile forces only. These links have one way rigidity and would be considered a link when is tension but not under compression.
4. **Fluid link** can transmit motion by pressure or compression only. Examples: hydraulic press, jacks, brakes, etc.
5. Spring are not considered links as they have no effect on the kinematics of a device but do play a role in supplying forces. These force are neglected during kinematic analysis and are considered during dynamic analysis.

1.3.2 Classification of Kinematic Pairs

The links of a mechanism are connected together for transmission of motion. The connections or joints between the links are called kinematic pairs. The kinematics of a mechanism depends upon the following factors.

1. Distance between successive joints.
2. Nature of the joint.
3. Relative motion permitted by the joint.
4. Constraints to produce the proper relative motion for the particular task to be performed.

The kinematic pairs can be classified as follows.

1. Releaux Classification

Releux has divided the kinematic pairs into higher pairs and lower pairs.

(i) **Higher pairs**: The surface contact between the elemental surfaces has line or point contact.

Examples: mating gear teeth, a wheel rolling on a rail, a ball rolling on a flat surface, cam contacting its roller follower.

There is a special subcatagory of higher pairs called **wrapping pairs**. Examples: connection between a belt and a pulley, between a chain and sprocket, between a rope and a drum. In wrapping pairs, one of the links has one-way rigidity.

There are infinite number of higher pairs and a systematic accounting is not possible. Each will be treated as a separate situation.

(ii) **Lower pairs**: A pair of links having surface or area contact between the members is called lower pair. The contact surface of the two links are similar.

Examples: universal joint, pin joint, nut and bolt, shaft rotating in a bearing, all pairs of slider-crank mechanism, etc.

2. Classification of Lower Pairs by Hartenberg and Denavit

There are six lower pairs shown in Table 1.1 which gives the name of pair, symbol used, pair variables, degrees of freedom and relative motion.

Table 1.1 Relative Motion in Lower Pairs

Sl. No.	*Pair*	*Fig. 1.3*	*Symbol*	*Pair Variation*	*Degrees of freedom*	*Relative Motion*
1.	Revolute	*(a)*	*R*	$\Delta\theta$	1	Circular
2.	Prism	*(b)*	*P*	Δs	1	Rectilinear
3.	Screw	*(c)*	*S*	$\Delta\theta$ or Δs	1	Helical
4.	Cylinder	*(d)*	*C*	$\Delta\theta$ and Δs	2	Cylindric
5.	Sphere	*(e)*	*G*	$\Delta\theta$, $\Delta\phi$, $\Delta\psi$	3	Spheric
6.	Flat	*(f)*	*F*	Δx, Δy, $\Delta\theta$	3	Planar

The six lower pairs are illustrated in Fig. 1.3

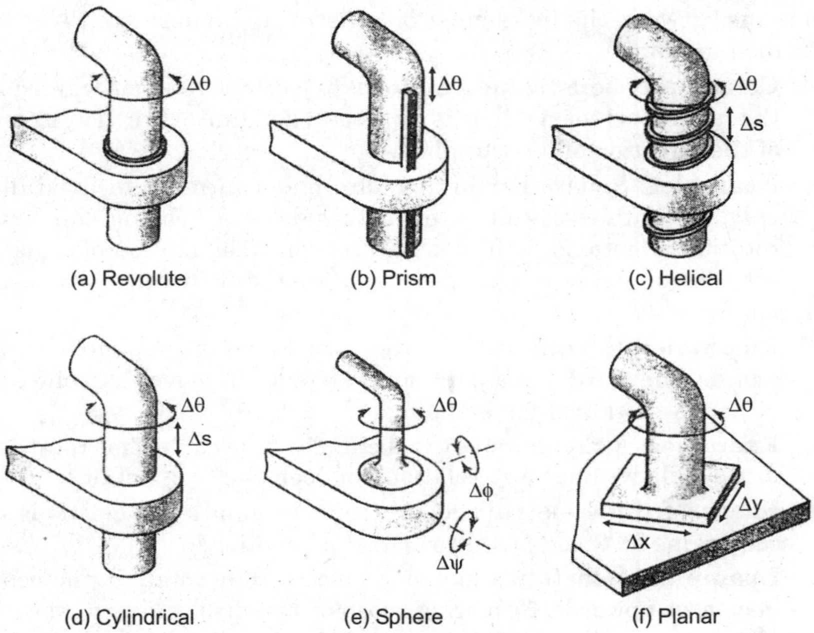

Fig. 1.3. Types of lower pairs

(*a*) **The turning pair or revolute:** It permits only relative rotation and hence has one degree of freedom. This pair is often called as pin joint.

Examples: A circular shaft rotating in a bearing. The connecting rod-crank pair.

(*b*) **The prismatic pair**: It permits only a relative sliding motion and therefore is often called a sliding joint. It also has a single degree of freedom.

Examples: A circular shaft with key joint sliding in a bore.

(*c*) **The screw pair or helical pair:** It permits a relative turning and sliding motion. The sliding and rotational motions are related by the helix angle of the thread and thus has one degree of freedom. The pair variable may be chosen either Δs or $\Delta\theta$ but not both.

Examples: the lead screw and nut of a lathe.

(*d*) **The cylindrical pair**: It permits both angular rotation and an independent sliding motion. It has two degrees of freedom.

(*e*) **The globular or spherical pair**: It is a ball-and-socket joint. It has three degrees of freedom, a rotation about each of the coordinate axes.

Examples: Ball and socket joint, universal joint.

(*f*) **The flat or planar pair:** It is rarely found in mechanisms. It has three degrees of freedom, rectilinear motion along each of the coordinate axes.

1.4. TYPES OF CONSTRAINTS

In order to produce the proper relative motion suitable for a particular task, the motions between the links must be constrained. There are three types of constrained motions.

1. **Completely constrained motion:** In a completely constrained motion, the motion between a pair is limited to a definite direction irrespective of the direction of force applied.

   Examples: Square bar in a square hole can move in axial direction only. A shaft with collars on both side of a bearing can have only rotational motions with no axial freedom. The motion of a piston in a cylinder of steam engine can make to-and-fro motion irrespective of motion of crank.

2. **Incompletely constrained motion:** In an incompletely contrained motion, the motion in a pair can take place in move than one direction under the action of force applied.

   Examples: A circular shaft in a circular hole can either rotate or slide in the hole without any relation between the two motions.

3. **Successfully constrained motion:** An additional means is used to constrains an incompletely constrained motion.

   Examples: A shaft in a foot-step bearing may rotate in the bearing or may move upwards. By placing a load on the shaft, the upward movement may be constrained. The motion of the pair is said to be successfully cobstrained. The motion of the valves of an I.C engines are successfully constrained by using a spring to keep the valves on the seats in closed position.

The kinematic pairs are also classified as follows.

1. **Closed pair**: The links of a closed kinematic pair are held together mechanically.
   Examples: All lower pairs.
2. **Unclosed pair:** The links are not held together mechanically in unclosed pair.
   Examples: Cam and follower are not held together mechanica lly but by gravity or spring force.

1.5. DEGREES OF FREEDOM OF PLANAR MECHANISMS

(1) Mechanisms may be categorized into following groups.

1. Planar mechanism
2. Spherical mechanism
3. Spatial mechanism.

The vast majority of mechanisms in use today are planar. The loci of all points are plane curves parallel to a single common plane. The motion transformation is called coplanar.

Examples: Plane four-bar linkage, the plate cam and follower, slider-crank mechanism. These mechanisms utilize only revolute and prismatic pairs. The graphical technique are well-suited to their solutions.

(2`) The number of degrees of freedom, also called **mobility** of the device is required for the design and analysis of a mechanism. The mobility of a mechanism can be determined directly from a count of the number of links and the number and types of joints in the mechanism. **Kutzback criterion** can be used to find the mobility or degrees of freedom of planar mechanisms.

(3) The mobility m of a planer n-link mechanisms is given by the following equation as per Kutzback criteria

$$m = 3(n-1) - 2j_1 - j_2$$

where j_1 = number of single-degree-of- freedom pairs

j_2 = number of two-degree-of-freedom pairs.

(4) Application of Kutzback criterion to planar mechanism.

(*a*)

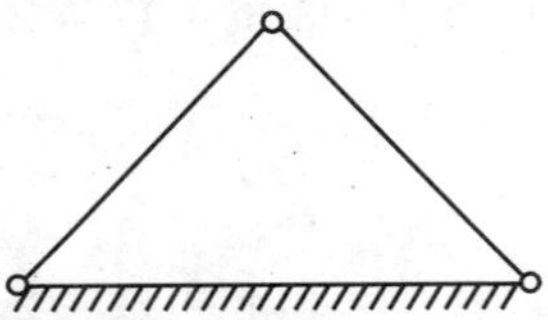

$n = 3,\ j_1 = 3,\ j_2 = 0$

$\therefore\ m = 3\ (n - 1) - 2j_1 - j_2$

$= 3\ (3 - 1) - 2 \times 3 - 0 = 0$

The motion is impossible. The mechanisms forms a structure.

(*b*)

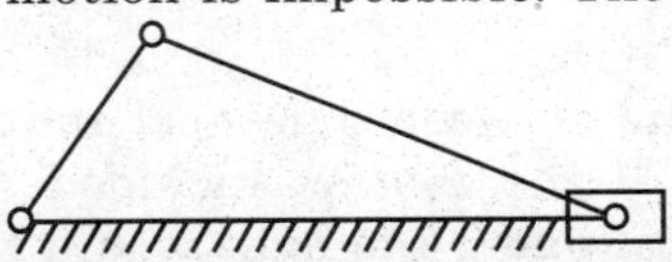

$n = 4,\ j_1 = 4,\ j_2 = 0$

$\therefore\ m = 3\ (n - 1) - j_1 - j_2$

$= 3\ (4 - 1) - 2 \times 4 - 0 = 1.$

The mechanisms can be driven by a single input motion.

(*c*)

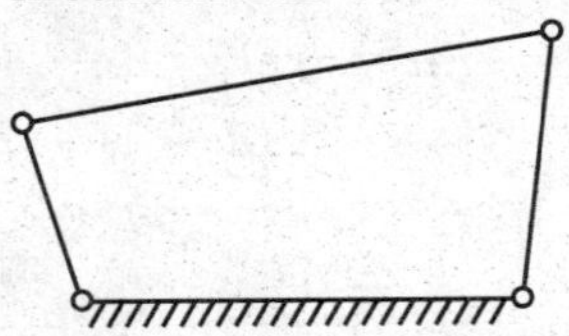

$n = 4,\ j_1 = 4,\ j_2 = 0$

$\therefore\ m = 3\ (4 - 1) - 2 \times 4 - 0 = 1.$

The mechanism can be driven by a single input motion.

(*d*)

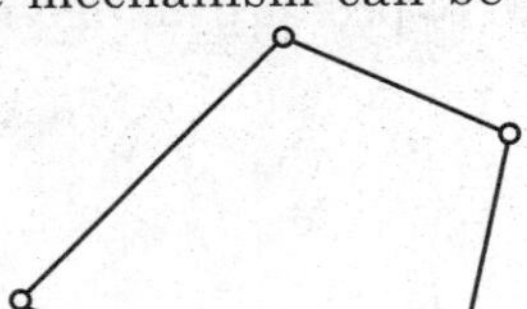

$n = 5,\ j_1 = 5,\ j_2 = 0$

$m = 3\ (n - 1) - 2j - j_2$

$= 3\ (5 - 1) - 2 \times 5 - 0 = 2$

Two separate input motions are necessary to produce constrained motion for the mechanism.

(*e*)

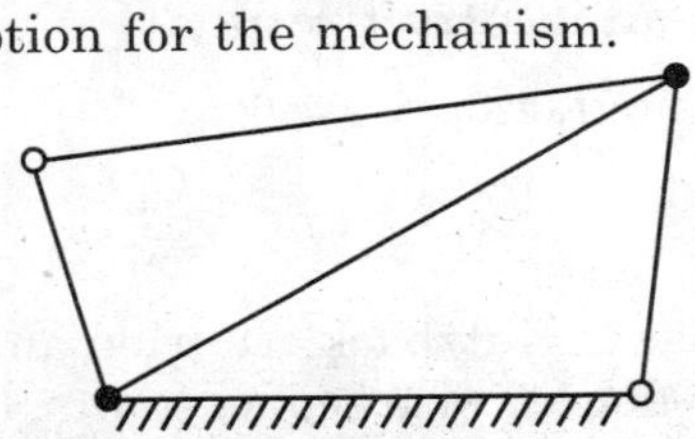

$n = 5,\ j_1 = 6,\ j2 = 21\ 0$

$m = 3(n - 1) - 2j1 - j2$

$= (5 - 1) - 2 \times 6 - 0 = 0$

The motion is impossible. The mechanism is a structure. Three links are joined by a single pin (•) two joints are counted. The connection is treated as two separate but concentric pairs.

(*f*)

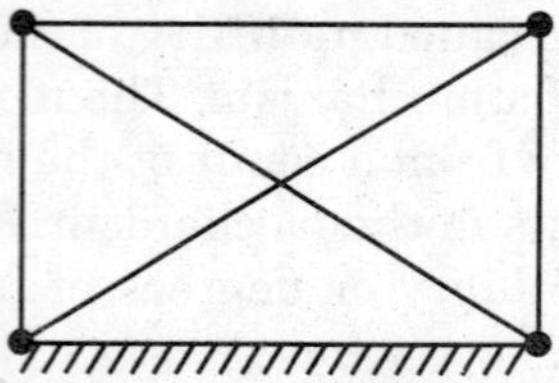

$n = 6, j_1 = 8, j_2 = 0$

$m = 3(n - 1) - 2j_1 - j_2$

$= 3(6 - 1) - 2 \times 8 - 0 = -1$

There are redundant constraints in the chain. It forms a statically indeterminate structure.

(*g*)

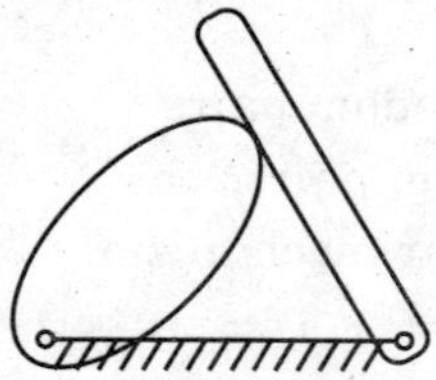

$n = 3,$

$j_1 = 2,$

$j_2 = 1$

$m = 3(n - 1) - 2j_1 - j_2$

$= 3(3 - 1) - 2 \times 2 = 1$

(*h*)

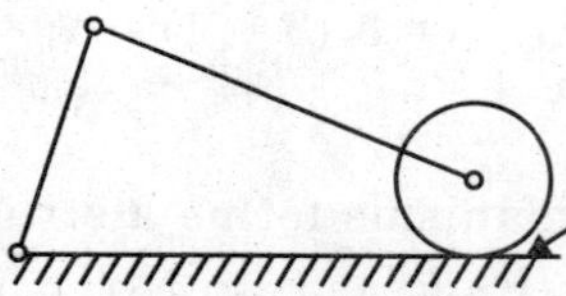

$n = 4, j_1 = 3, j_2 = 1$

$m = 3(n - 1) - 2j_1 - j_2$

$= 3(4 - 1) - 2 \times 3 - 1 = 2$

The two-degree of freedom joint between the wheel and fixed link has slipping between the links. If the contact included gear teeth or friction was high to prevent slipping the joint would be counted as one-degree-of-freedom pair.

(*i*)

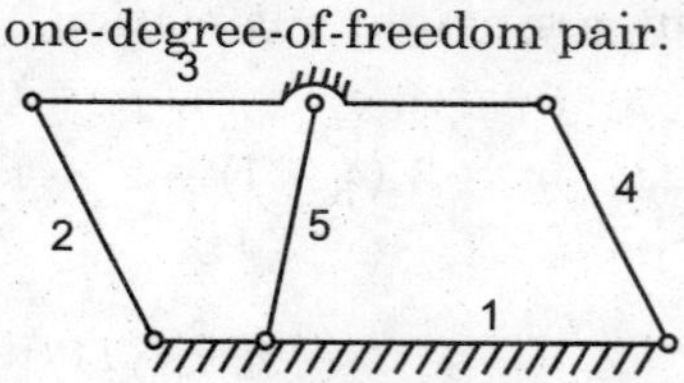

$n = 5, j_1 = 6, j_2 = 0$

$\therefore m = 3(n - 1) - 2j_1 - j_2$

$= 3(5 - 1) - 2 \times 6 - 0$

$= 0$

It is a structure. If link 5 is arranged to produce double-parallelogram linkage as shown in Fig. (*j*).

(*j*)

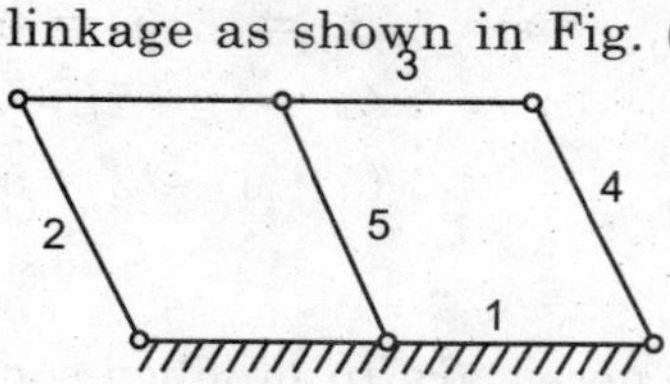

$n = 5,$

$j_1 = 6,$

$j_2 = 0$

$m = 1$

The Kutzback criterion gives an incorrect result.

Fig. 1.4. Applications of Kutzback criterion

1.6. GRUBLER'S EQUATION

The Grubler's criterion is a special case of Kutzback criterion and applies to planar mechanisms with only single-degree of freedom joints. The overall mobility of the mechanism is unity

Put $m = 1$ and $j_2 = 0$ in Kutzback equation, we get

$$1 = 3(n - 1) - 2j_1$$

or, $$3n - 2j_1 - 4 = 0$$

This equation is called Grubler's equation for planar mechanisms with constrained motion.

The planar mechanisms with mobility of 1 and single-degree-of-freedom joints will have number of links (n) an even number.

Examples: The simplest mechanisms of this type are :

1. Four-bar mechanism
2. Slider-crank mechanism

where $n = 4$ and $j_1 = 4$

1.7. LINKAGE MECHANISMS

There are a wide variety of mechanisms including both higher pairs and lower pairs. A linkage mechanism has only lower pairs. A linkage is therefore, connected only by lower pairs as per Fig. 1.3

A mechanism can be obtained by fixing one link in a kinematic chain. Therefore, as many mechanisms as the number of links in a kinematic chain can be obtained by fixing in turn, different links in a kinematic chain. This method of obtaining different mechanisms by fixing different links in a kinematic chain is known as **inversion of the mechanism.** There is no change in the relative motions between the various links through the process of inversion. However, the absolute motions of the links with respect to fixed link may change drastically.

The part of the mechanism which initially moves with respect to the fixed link or frame is called **driver** and the part of the mechanism to which motion is transmitted is called **follower**.

Most of the mechanisms are reversible and the same link can play the role of a driver and follower. For example in an I.C. engine, the piston is the driver and crank is the follower. In a reciprocating air compressor, the crank becomes the driver and piston the follower.

1.7.1. Types of Linkage Mechanisms

Many attempts have been made in devising a method of classification of linkage mechanisms but none have been very successful. Torfason has classified the mechanisms according to type of motion transformation. He has displayed 262 mechanisms.

The most important kinematic chains consist of four lower pairs, each pair having a sliding pair or a turning pair. The following three types of kinematic chains are discussed in details.

1. Four-bar chain or quadric cycle chain.
2. Single slider-crank chain.
3. Double slider-crank chain.

1.7.2. Four-Bar Chain or Quadric Cycle Chain

A four-bar chain is the most fundamental of the planar kinematic chains. This mechanical device can be used for the mechanization and control of motion due to its simplicity and versatility. Basically it consists of four rigid links connected in the form of a quadrilateral by four pair joints. When one of the links is fixed, it is known as linkage mechanism. A link that makes the complete revolution is called the crank, the link opposite to the fixed link, the **coupler** and the forth link a **lever or rocker** if it oscillates or another crank, if it rotates.

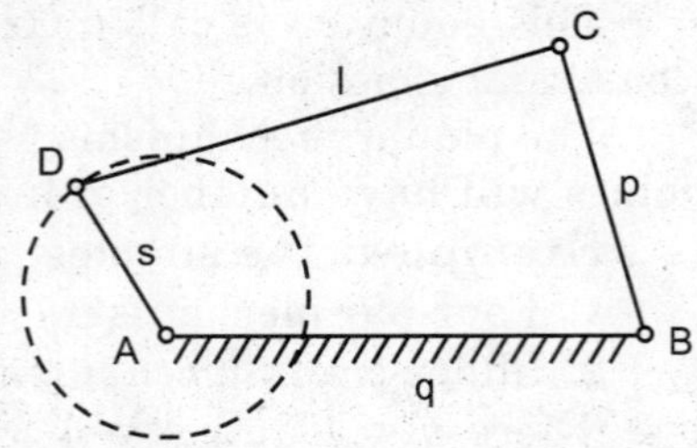

Fig. 1.5. Four-bar chain

The four links may be of different lengths. According to **Grashof's law "for a planar four-bar linkage, the sum of the shortest and the longest link lengths cannot be greater than the sum of the remaining two link lengths if there is to be continuous relative rotation between two member"**. In the above figure the shortest link AD will rotate continuously relative to the other three links if and only if

$$s + l \leq p + q$$

where s = length of shortest link

l = length of longest link.

p and q = lengths of other two links

The following observation may be made from Grashof 's law

1. No link will make a complete revolution relative to another if Grashof 's law is not satisfied.
2. A linkage in which no link makes a complete revolution will not be useful.
3. The law does not specify the order in which the links of various lengths are to be connected.
4. The law does not specify which link showed be fixed. Any of the four links can be fixed.
5. Different inversions can be obtained by fixing different links and locating the shortest link with respect to other links.

1.7.3. Single Slider-Crank Chain

A single slider-crank chain is a modification of the basic four-bar chain. The four-bar chain has four links each forming a turning pair *A, B, C* and *D* (Fig. 1.5). The single-slider crank chain has one sliding pair and three turning pairs. It is found in all reciprocating machines such as steam engine, I.C. engines, reciprocating pumps and compressors. This type of linkage mechanism converts reciprocating motion into rotary motion and vice-verse.

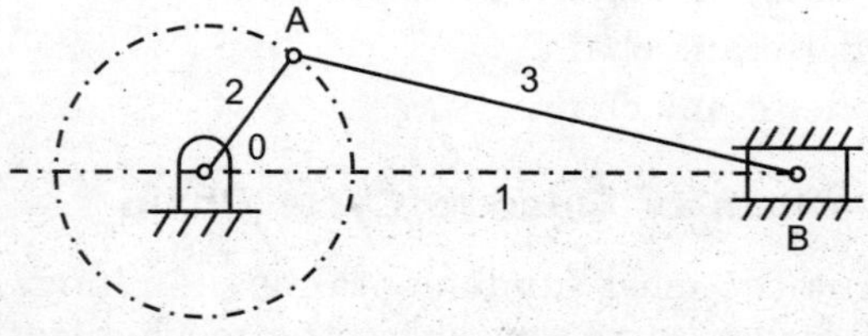

Fig. 1.6. Single slider-crank chain

The link 1 corresponds to the frame which is fixed. The link 2 corresponds to the crank and link 3 corresponds to connecting rod. As the crank rotates, the piston reciprocates in the cylinder.

In a slider-crank chain, the straight line path of the slider (piston) may be passing though the fixed pivot 0 (center line of crank shaft). If it is displaced by a distance the same is called offset and the chain formed is called **off-set slider-crank chain**.

A slider-crank chain has many inversions obtained by fixing different links of the kinematic chain.

1.7.4. Double Slider-Crank Chain

A four-bar chain consisting of two turning pairs and two sliding pairs is known as double slider-crank chain. In the Fig. 1.7, link 2 and link 1 form one turning pair and link 2 and 3 form second turning pair. The link 3 and link 4 form one sliding pair and link 1 and link 4 from the second sliding pair.

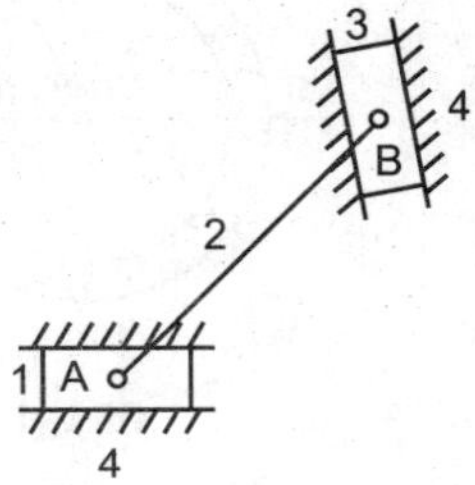

Fig 1.7 Double slider-crank chain

1.8. INVERSIONS OF FOUR-BAR CHAIN

There are many inversions of four-bar chain. The following inversions are important from practical point of view.

1. Beam engine, i.e., crank and lever mechanism.
2. Compiling rod of a locomotive, i.e., double crank mechanism.
3. Watt's indicator mechanism, i.e., double lever mechanism.

1.8.1. Beam Engine

It is used to convert rotary motion into reciprocating motion. A beam engine is shown in Fig. 1.8. It has four links.

1. Link 1 is fixed frame.
2. Link 2 is crank rotating about fixed centre 0.
3. Link 3 is connecting rod.
4. Link 4 is lever oscillating about fulcrum *F*.

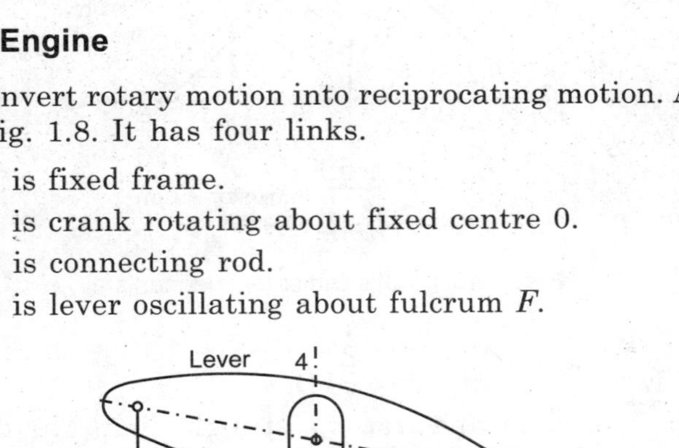

Fig. 1.8. Bean engine

When the crank rotates, the connecting rod makes the lever oscillate. The piston is pushed up and down in the cylinder.

1.8.2. Coupling Rod of a Locomotive

The mechanism is used to transmit rotary motion from one wheel of locomotive to another wheel. The mechanism is shown in Fig. 1.9. It has four links.

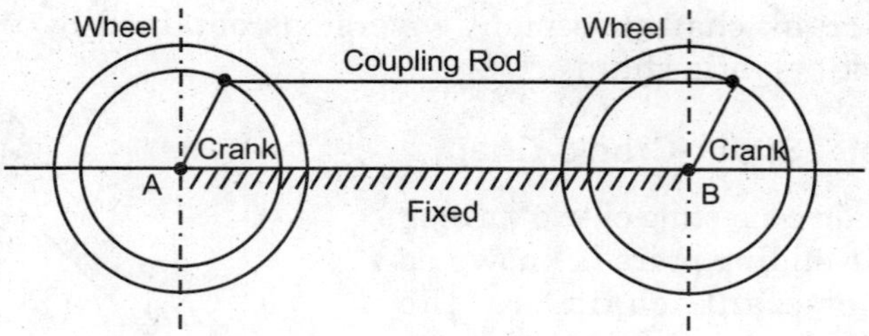

Fig. 1.9. Coupling rod of a locomotive

1. Fixed link to maintain constant distance between wheel axles.
2. Crank of equal length connected to wheels.
3. Coupling rod.

The rotary motion of wheel *A* is transmitted to wheel *B*.

1.8.3. Watt's Indicator Mechanism

The mechanism is used to plot indicator diagram of piston cylinder engine. The mechanism is shown in Fig. 1.10. It has four links.

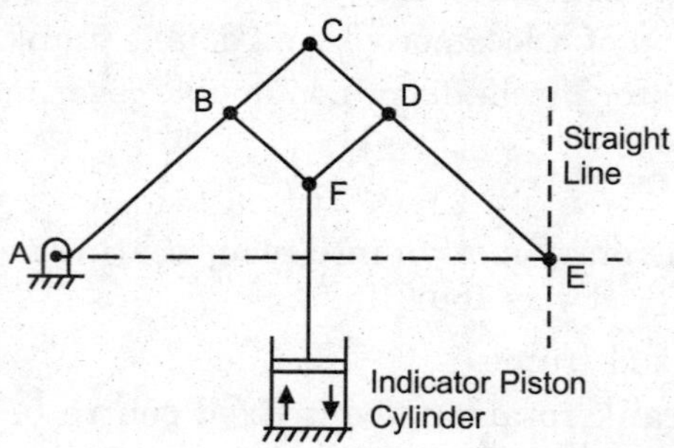

Fig. 1.10. Watt's indicator mechanism

1. Fixed link A.
2. Link AC
3. Link CE
4. Link BFD

E is the tracing point and traces a straight link. The displacement of link BFD is proportional to fluid pressure in the cylinder which results in the motion of point E.

1.9. INVERSIONS OF SINGLE SLIDER CRANK CHAIN

The single slider crank chain is a four-link mechanism and therefore four mechanism can be obtained by fixing different links one by one. The following inversions have found practical applications.

1. Pendulum Pump or Bull Engine.
2. Oscillating Cylinder Engine.
3. Rotary Internal Combustion Engine (Grome Engine).
4. Quick Return Mechanism (Crank and Slotted Lever).
5. Whitworth Quick Return Motion Mechanism.

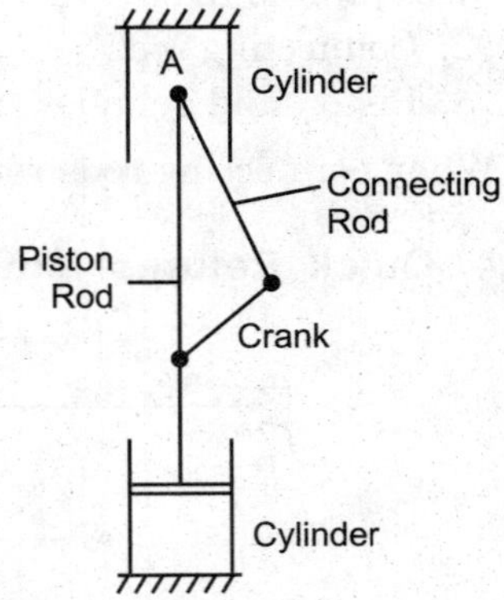

Fig. 1.11. Pendulum pump

1.9.1. Pendulum Pump or Bull Engine

A pendulum pump is used as feed water pump for boilers. A schematic diagram of such a pump is shown in Fig 1.11. It consists of the following links.

1. Cylinders fixed
2. Crank
3. Connecting rod

The crank rotates and connecting rod oscillates about pivot A attached to cylinder. The piston attached to piston rod reciprocates and the piston pumps water.

1.9.2. Oscillating Cylinder Engine

This mechanism is used to convert reciprocating motion into rotary motion. The mechanism is shown in Fig. 1.12 and has the following links.

1. The connecting rod is fixed
2. Crank
3. Piston rod
4. Cylinder.

When the crank rotates, the piston rod and piston reciprocate and cylinder oscillates about pivot *A*.

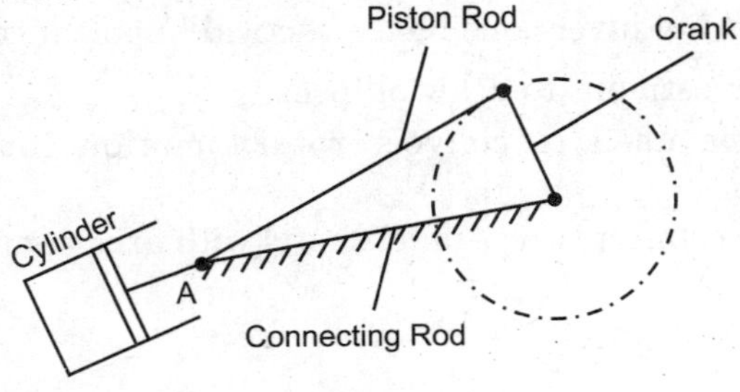

Fig. 1.12 Oscillating cylinder engine

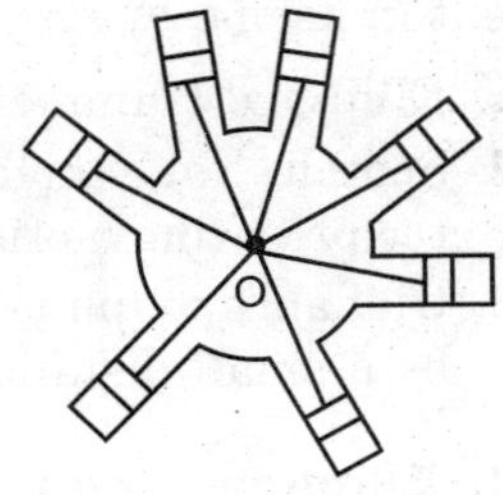

Fig. 1.13 Rotary Combustion engine

1.9.3 Grome Engine or Rotary Combustion Engine

It is used as aviation engine. The arrangement of mechanism with seven cylinder is shown in Fig. 1.13. It has following links.

1. Crank is fixed
2. Connecting rods
3. Piston and cylinder

When connecting rods rotate, the piston reciprocates inside the cylinders.

1.9.4. Quick Returns Motion with Crank and Slotted Lever

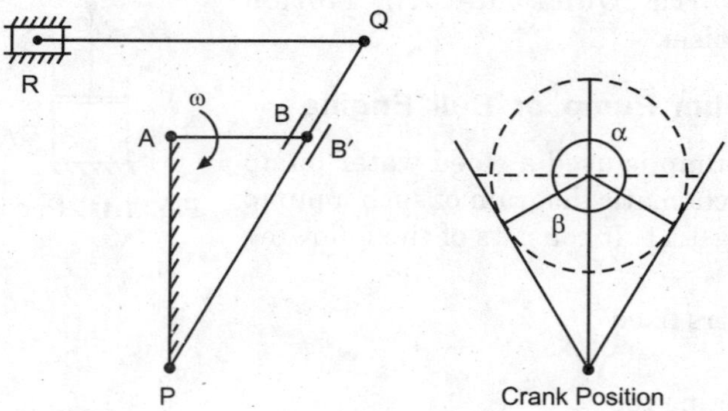

Fig. 1.14 Quick returns motion with crank and slotted lever

It has fixed link PA and crank AB rotates about point A. The slotted lever PQ has a slider B. At the end of link QR, tool box R is attached to move to and fro.

The two extreme position of the crank are shown.

$$\frac{\text{Time of cutting stroke}}{\text{Time of return stroke}} = \frac{\alpha}{\beta} = \frac{\alpha}{(360^\circ - \alpha)}$$

Angle α is move than angle β; the cutting tme will be more than return time.

1.10. INVERSIONS OF DOUBLE SLIDER CRANK CHAIN

Different inversions of double slider crank chain can be obtained by fixing different links one-by-one. The following inversions have practical importance.

1. Elliptical trammels is an investment to draw ellipses.
2. Scotch Yoke mechanism is used to convert rotary motion into reciprocating motion.
3. Oldham's coupling is used to connect two parallel shafts with axes apart by a small distance.

1.10.1. Elliptical Trammels

There are two sliders 1 and 3 interconnected by a link 2. These sliders are free to move inside the slotted frame 4. The slider 1 and 3 move in the fixed slots. Any point C on the extension of the link AB traces an ellipse.

The semi-major axis of ellipse.

$$AC = b$$

The semi-minor axis of ellipse

$$BC = a$$

Therefore the coordinates of C be x, y.

$$\frac{x}{BC} = \cos\theta$$

$$\frac{x}{a} = \cos\theta$$

and

$$\frac{y}{AC} = \sin\theta$$

$$\frac{y}{b} = \sin\theta$$

$$\therefore \quad \frac{x^2}{a^2}+\frac{y^2}{b^2} = \sin^2\theta + \cos^2\theta = 1$$

This is the equitation of an ellipse.

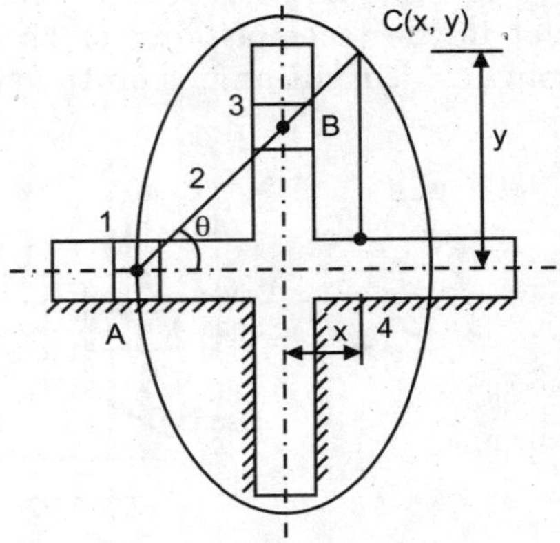

Fig. 1.15 Elliptical trammel.

1.10.2. Scotch Yoke Mechanism

The mechanism consists of the following links :

Links 1: Fixed

Links 2: It makes two turning pairs with link 3 and 1.

Links 3: Sliding link

Links 4: Reciprocating link.

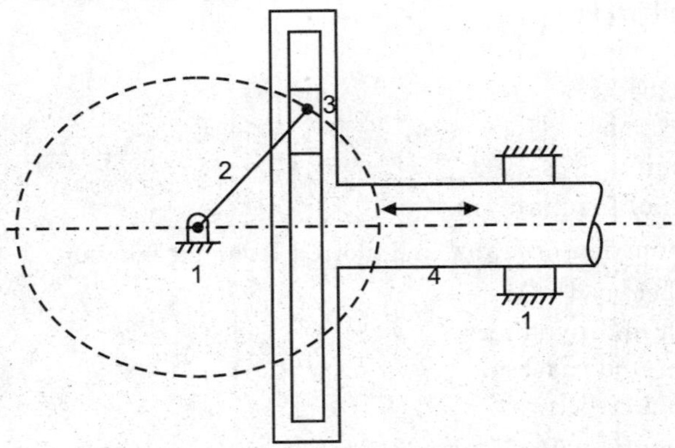

Fig. 1.16. Scotch Yoke Mechanism.

The rotary motion of link 2 (crank) is converted into reciprocating motion of links 4 (piston of a compressor or a pump).

1.10.3. Oldham's Coupling

The coupling has the following links :

Link 1 is the driven shaft B with coupling flange C.

Link 2 is the fixed frame on which the bearings of the two shafts are supported.

Link 3 is the driving shaft A with coupling flange D.

The coupling flanges fit together with the help of tongue and diametrial slot in the faces of flanges. There is an intermediate flange E which has tongues at right angles on the two opposite faces.

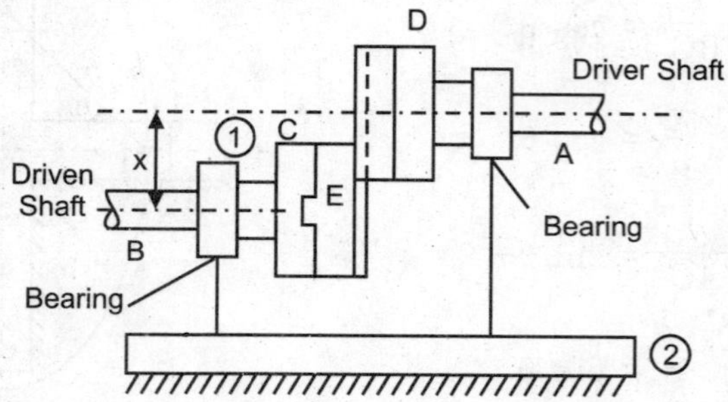

Fig. 1.17. Oldham's coupling.

The maximum sliding speed of each tongue (on E) along slots (C and D)

$$v = x \times \omega$$

where ω = angular velocity of each shaft (rad/s).

v = linear velocity (m/s).

x = distance between axis of the two shafts.

QUESTION BANK

1. Define the following.
   (*i*) Link
   (*ii*) Resistant body
   (*iii*) Kinematic pair
   (*iv*) Mechanism
   (*v*) Inversion of mechanism. **[UPTU 2010–11]**
2. Describe the function of crank and slotted lever mechanism. **[UPTU 2010-11]**
3. Differentiate between
   (*a*) Kinematics and dynamics
   (*b*) Mechanism and machine
   (*c*) Machine and structure.
4. How are kinematic pair classified? Explain with examples.
5. Explain different types of links.
6. What is degree of freedom of a mechanism? How is it found out ?
7. Explain different types of constraints.
8. What is Kutzback critieria? How is degree of freedom found out?
9. What is the difference between Kutzback criterion and Grubler's criterion ?
10. Describe various inversions of a slider-crank mechanism giving examples.
11. Enumerate the inversions of a double slider-crank chain. Give examples.

CHAPTER 2

VELOCITY ANALYSIS IN MECHANISMS

2.1. INTRODUCTION

The theory of machines is concerned with the motion of the parts of machines and the forces acting on these parts. Each point of every link of a machine is constrained to move along a definite path. The points on certain links move along straight lines or circular arcs or non-circular curved paths.

For example, all points on piston, piston rod, crosshead of a reciprocating engine move along parallel straight lines. All points on the crank and crankpin move along circular arcs about axis of crankshaft. However, any point on the connecting rod moves along non-circular curved paths whose radius of curvature changes from time to time.

Displacement as well as velocity are vector quantities. The velocity will be along the direction of displacement for liner motion and along tangent to the path of the point for circular and curved motion.

The velocities in machines and mechanisms can be determined by many methods.

1. Analytical Method

(*i*) Vector algebra (*ii*) Complex numbers.

These method are a bit tedious. However, with the availability of computers and calculators, use of analytical method are becoming popular.

2. Graphical Methods

Graphical methods of velocity analysis in mechanisms are very popular. The main advantages of graphical methods are achievement of quick solution, visualization of problem and insight into the problem. It gives accuracy to an acceptable degree.

There are two graphical methods.

(*i*) **Relative velocity method:** It can be used for any configuration of mechanism.

(*ii*) **Instantaneous centre method:** It can be used for simple mechanisms.

Both the graphical methods discussed are discussed alongwith their applications for specific mechanisms such as velocities in four-bar mechanisms, slider crank mechanisms and quick return motion mechanisms.

2.2. VELOCITY OF POINT IN MECHANISM

A link of mechanism may move in any direction in space. Plane motion is the motion of a body in single plane. Most of the mechanisms can be assumed to move in single plane.

Plane motion may be classified as follows:

1. Translation
2. Rotation
3. General plane motion.

2.2.1. Translation

If a straight line drawn on the moving link remains parallel to its original position at any time, the motion is called translation. If the path traced by any point on the link is a straight line, the motion is called **rectilinear translation.** If the path traced by the point is a curve, the motion is called **curvilinear translation.** Rectilinear translation is also known as linear motion and curvilinear translation as curvilinear motion. The motion of a point on a piston or piston rod of a reciprocating engine undergoes linear motion.

Different types of linear motion problem may be:

(i) Motion with uniform velocity

(ii) Motion with uniform acceleration

(iii) Motion under acceleration due to gravity

(iv) Motion with varying acceleration.

Many kinematic problems can be solved in linear motion by using the definition of velocity and acceleration. A body in motion changes its position relative to the surrounding. **Velocity** is the rate of change of displacement(s) with respect to surroundings in a particular direction. Velocity is a vector quantity.

$$v = \frac{ds}{dt}$$

Acceleration is the rate of change of velocity. The rate of change of velocity may be uniform and the motion is said to be with uniform acceleration. The link may move with variable acceleration if the rate of change of velocity is not uniform.

$$a = \frac{dv}{dt}$$

If the motion is with uniform velocity.

$$\frac{ds}{dt} = v$$

$$\therefore \quad s = \int v\,dt = vt \qquad (\because v = \text{constant})$$

2.2.2. Rotation

If all particles of a link move in concentric circles the motion is called rotation. The displacement takes place along a circular path. There is change of direction as well as change of magnitude of displacement. The velocity of the point on a rotating link is tangential to the curved path.

Where a point on the link moves from position A to B, the angular displacement θ is measured in radians. The rate of change of displacement with time is called **angular velocity** (ω).

$$\omega = \frac{d\theta}{dt}$$

The rate of change of angular velocity with time is called angular acceleration (α).

$$\alpha = \frac{d\omega}{dt} = \frac{d^2\theta}{dt^2}$$

Also $$\alpha = \frac{d\omega}{dt} = \frac{d\omega}{d\theta} \times \frac{d\theta}{dt} = \omega . \frac{d\omega}{d\theta}$$

For motion with uniform angular velocity, the angular displacement

$$\theta = \omega t \text{ radians.}$$

$$\alpha = 0$$

$$\omega = \frac{2\pi N}{60} \text{ red/s}$$

where N = No. of revolutions per minute (rpm).

The time taken for one revolution ($\theta = 2\pi$)

$$2\pi = \omega t$$

or $$T = \frac{2\pi}{\omega}$$

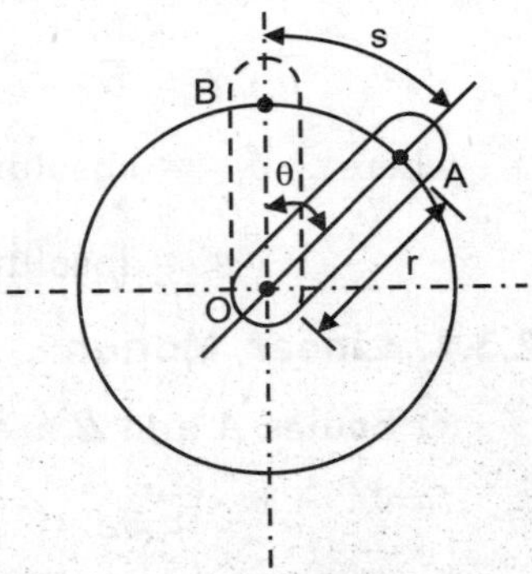

Fig. 2.1 Rotation of a link

From Fig. 2.1

$$s = r\theta$$

The tangential velocity of the particle is called **linear velocity** (ν)

$$\nu = \frac{ds}{dt} = r\frac{d\theta}{dt} = r\omega$$

2.2.3. General Plane Motion

Combination of both translation and rotation is called general plane motion. The example is motion of connecting rod of a reciprocating engine. The general plane motion of a point on a link is split into translation and rotary motion as shown in Fig. 2.2. The solution of translation motion and rotary motion are combined to get the solution of general plane motion.

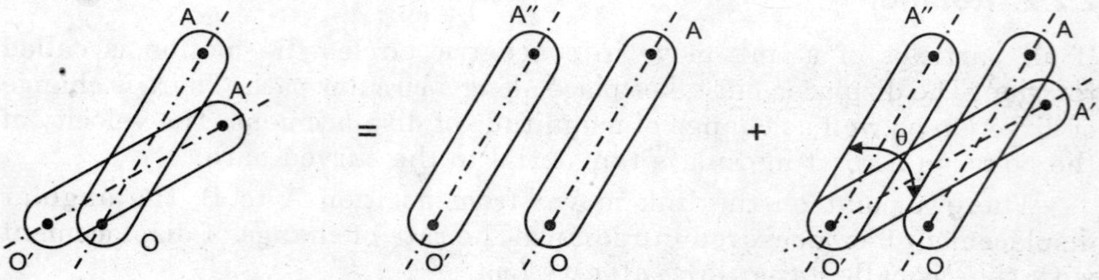

Plane motion = Translation + Rotation

Fig. 2.2 General plane motion

Two graphical methods will be discussed to find the velocity of a point in mechanism. The methods to find the acceleration of the point will be discussed in the next chapter.

2.3. RELATIVE VELOCITY METHOD

If a body (A) is morning will a velocity v_A and body (B) is moving with a velocity v_B with respect to stationary earth, then the velocity of moving body. A with respect to moving body B is called relative velocity of body A with respect to body B. It is the vector difference of absolute velocities v_A and v_B.

$$\vec{v}_{A/B} = \vec{v}_A - \vec{v}_B$$

where $\vec{v}_A$ = absolute velocity of A with respect to stationary earth.

$\vec{v}_B$ = absolute velocity of B with respect to stationary earth.

2.3.1. Linear Motion

(*i*) Bodies A and B are moving in the same parallel directions.

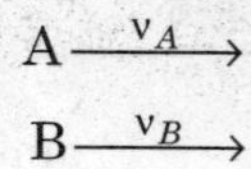

$$\vec{v}_{A/B} = \vec{v}_A - \vec{v}_B$$

If $\vec{v}_A$ = 70 m/s and

$\vec{v}_B$ = 10 m/s

$$\vec{v}_{A/B} = 70 - 10$$
$$= 60 \text{ m/s}$$

(*ii*) Bodies A and B moving in the opposite parallel direction.

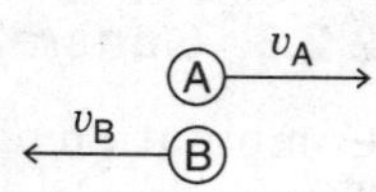

$$\vec{v}_{A/B} = 70 - (-10)$$
$$= 80 \text{ m/s}$$

2.3.2 Plane Motion

In order to find relative velocity in plane motion, find the components of velocities in X and Y direction.

$$v_{A/B_x} = v_{A_x} - v_{B_x}$$

$$v_{A/B_y} = v_{A_y} - b_{B_y}$$

$$v_{A/B} = \sqrt{(v_{A/B_x})^2 + (v_{A/B_y})^2}$$

$$\tan \alpha = \frac{v_{A/B_y}}{v_{A/B_x}}$$

Where α is the direction of relative velocity ($v_{A/B}$) with respect to x-axis.

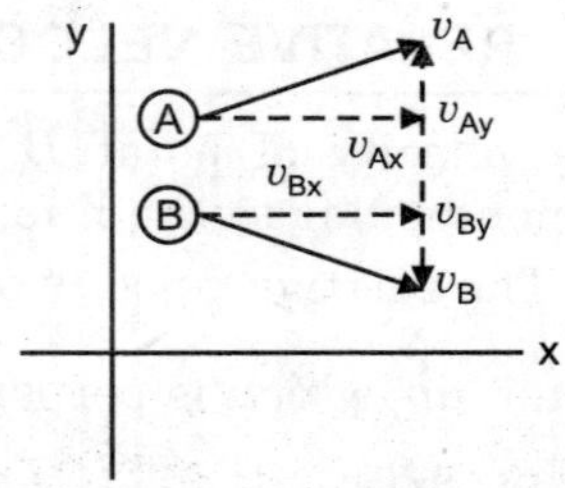

Fig. 2.3 Plane motion

2.3.3. Relative Distance

The two bodies are moving with velocities v_A and v_B in a plane. The distance between the two bodies is AB. In order to find relative distance between A and B after time interval t, follow procedure as follows.

(*i*) Assure body B is stationary

(*ii*) Move body A with relative velocity $v_{A/B}$.

(*iii*) Calculate distance travelled by A as $v_{A/B} \times t$ and move body A by the above distance to position A'.

(*iv*) Join $A'B$. This is the relative distance between bodies A and B after time interval t.

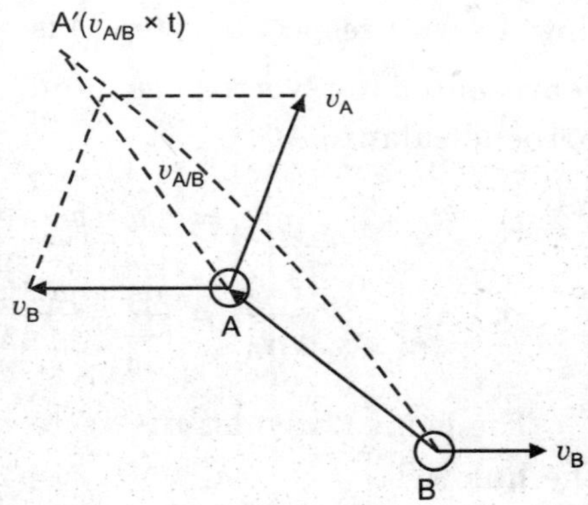

Fig. 2.4. Relative Distance

The alone procedure is illustrated in Fig. 2.4.

2.3.4. Relative Velocity

A body is moving with a velocity v_1. It is acted upon by an external force which imparts the body a velocity v_2 in the direction of force. The combined effect of velocities v_1 and v_2 is the relative velocity (v).

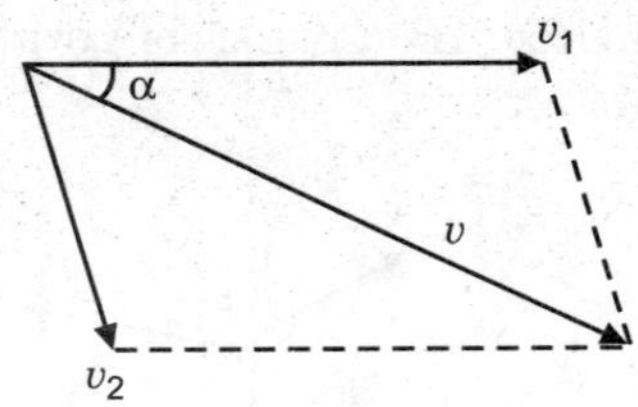

Fig. 2.5. Result and velocity

2.3.5. Applications

The relative velocity method is a universal graphical method to find the velocities in any configuration of mechanism. Its specific applications for the following mechanisms are of practical importance.

1. Four-bar mechanisms
2. Slider crank mechanisms
3. Quick return motion mechanisms
4. Rubbing velocity at a pin joint.

2.4. RELATIVE VELOCITY OF A RIGID LINK

The velocity of point B with respect to point A on a rigid link AB is perpendicular to the line AB.

The relative velocity of point B with repsect to A ($v_{A/B}$) is represented by vector $\overrightarrow{ab}$ which is perpendicular to line AB.

$$v_{B/A} = \overrightarrow{ab} = \omega \times AB$$

where ω= angular velocity of link AB about A.

The velocity of any point C on link AB with respect to A ($v_{C/A}$) is represented by vector $\overrightarrow{ac}$ and perpendicular to AC.

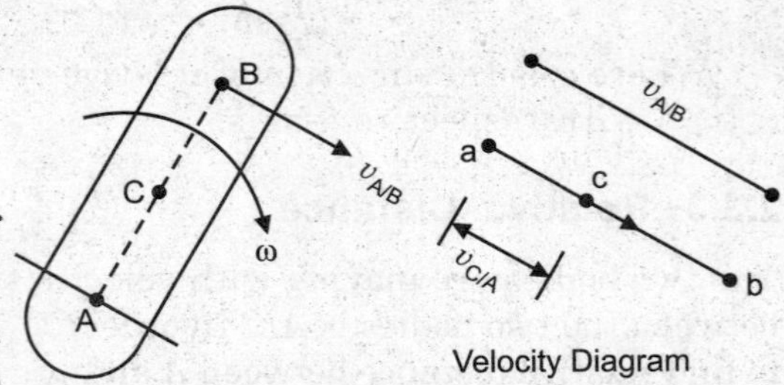

Fig 2.6. Relative velocity of a link

$$\therefore \quad v_{C/A} = \overrightarrow{ac} = \omega \times AC$$

$$\therefore \quad \frac{v_{C/A}}{v_{B/A}} = \frac{\overrightarrow{ac}}{\overrightarrow{ab}} = \frac{\omega . AC}{\omega . AB} = \frac{AC}{AB}$$

The point C divides the vector $\overrightarrow{ab}$ in the same ratio as point C divides the link AB.

2.5 VELOCITIES IN FOUR-BAR MECHANISM

A four bar mechanism $ABCD$ is show in Fig 2.7 in which link AD is fixed and link AB (crank) is rotating anti-clockwise with an angular velocity ω (known).

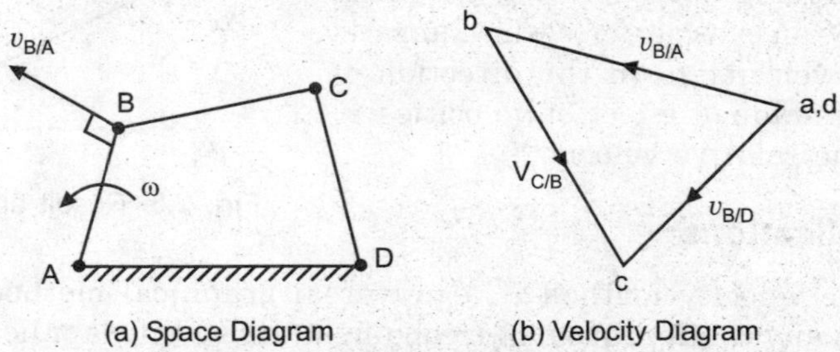

Fig 2.7. Velocities in four bar mechanism

The linear velocity of point B with respect to point A.

$$v_{B/A} = \omega \times AB$$

The point A is fixed.

$$v_A = 0$$
$$v_{B/A} = v_B$$
$$v_{D/A} = 0$$

The velocity $v_{B/A}$ will be perpendicular to link *AB*.

The velocity diagram can be constructed as follows.

1. Select a suitable scale and plot point *a*. Draw vector $\overrightarrow{ab}$ perpendicular to link AB so that $\overrightarrow{ab}$ represents $v_{B/A}$.
2. The point *d* coincides with point *a* as link *AD* is fixed.
3. The velocity of point *C* with respect to point *B* is represented by vector $\overrightarrow{bc}$ which is drawn perpendicular to link *BC* from point *b* to *c* in the vector diagram.
4. From point *a* (or *d*) draw vector $\overrightarrow{dc}$ perpendicular to link *DC* to represent velocity of point *C* with respect to point *D*. The vectors $\overrightarrow{bc}$ and $\overrightarrow{dc}$ intersect at *C*.
5. Δabc represents the velocity diagram of four-bar mechanisms *ABCD*.

   The angular velocity of link $BC = \dfrac{\overrightarrow{bc}}{BC}$

   The angular velocity of link $CD = \dfrac{\overrightarrow{cd}}{CD}$

Note: Use small letters for vector diagram and capital letters for space diagram.

Example 2.1. **In a four-bar mechanism link *AD* is fixed. The crank *AB* rotates clock wire at 120 rpm. Link *CD* oscillates about *D*. Find the angular velocity of link *CD* for the following data :**

$$AD = 15 \text{ cm}, AB = 4 \text{ cm}, CD = 8 \text{ cm}, BC = AD, \angle BAD = 60°$$

Solution: Draw space diagram to a suitable scale.

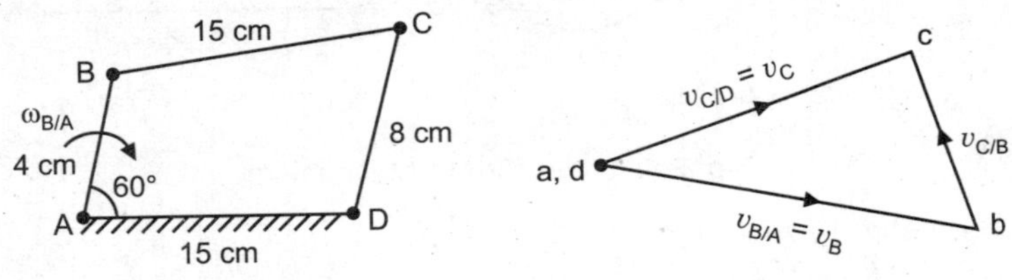

(a) Space Diagram (b) Velocity Diagram

Angular velocity of link *AB*,

$$\omega_{AB} = \frac{2\pi N_{AB}}{60} = \frac{2\pi \times 120}{60} = 12.56 \text{ rad/s.}$$

$$v_{B/A} = v_B = \omega_{AB} \times AB = 12.56 \times 4 = 50.24 \text{ cm/s.}$$

The direction of $v_{B/A}$ is perpendicular to link *AB*.

Construction of vector diagram *abc*.

1. Select a suitable scale. From any point a draw $\overrightarrow{ab}$ perpendicular to *AB* to represent $v_{B/A}$.

2. Link AD being fixed, point a and d coincide.
3. $v_{C/D}$ is perpendicular to link CD and $v_{C/B}$ is perpendicular to link CB.

   From point b draw $\overrightarrow{bc}$ perpendicular to BC.

   From point a or d draw $\overrightarrow{dc}$ perpendicular to DC.

   The vectors $\overrightarrow{bc}$ and $\overrightarrow{dc}$ interact at c.
4. $\Delta\, abc$ is the vector diagram of four-bar mechanism $ABCD$.
5. Measure $\overrightarrow{dc}$ = 38 cm/s.

   $\therefore \quad v_{C/D} = 38$ cm/s.

   Angular velocity of link CD.

$$\omega_{CD} = \frac{v_{C/D}}{CD} = \frac{38}{8} = 4.75 \text{ rad/s clockwise. } \textbf{Ans.}$$

2.6. VELOCITIES IN SLIDER CRANK MECHANISM

Crank AB of slider crank mechanism ABC is rotating with a known speed ω in the clockwise direction. The slider C more along line CA. The connecting CB oscillates in the space.

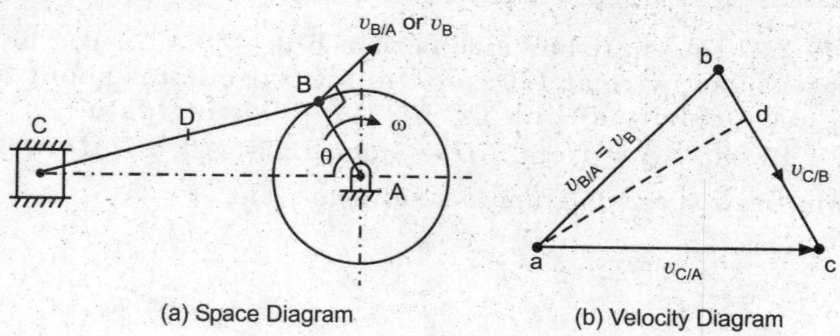

Fig 2.8. Slider Crank Mechanism

Angular velocity of crank BA

$$\omega = \frac{2\pi N}{60}$$

where N = Speed of crank BA (rpm).

The point A is fixed.

$\therefore \quad v_{B/A} = v_B = \omega \times AB$

This velocity is perpendicular to link AB.

Construction of Velocity Diagram

1. Select a suitable scale and draw $\overrightarrow{ab}$ from any point a and perpendicular to AB.

Velocity of point B w.r.t. point A,

$$v_{B/A} = v_B = \omega \times AB = \overrightarrow{ab}$$

2. From point b draw $\overrightarrow{bc}$ perpendicular to BC to represent $v_{C/B}$.
3. From point a draw $\overrightarrow{ac}$ horizontally to represent $v_{C/A}$.
4. The vectors $\overrightarrow{bc}$ and $\overrightarrow{ac}$ interest at point c.
5. $\Delta\, abc$ is the velocity diagram of slider crank mechanism.
6. The absolute velocity of any point D on connecting rod BC will be given by $\overrightarrow{bd}$ such that

$$\frac{\overrightarrow{bd}}{\overrightarrow{bc}} = \frac{BD}{BC}$$

or
$$\overrightarrow{bd} = \frac{BD}{BC} \times \overrightarrow{bc}$$

Join point a to d. The vector $\overrightarrow{ad}$ gives magnitude and direction of absolute velocity of point D.

Notes:

1. Use small letters for velocity diagram and capital letters for space diagram.
2. The absolute velocity are measured from fixed points.
3. Start the velocity diagram from a link whose velocity is known in magnitude and direction.

Example 2.2. For the position of the mechanism shown, calculate the angular velocity of the link *AR* where is *OA* 300 mm and rotates at 20 rad/s in the clockwise direction. [UPTU 2010-11]

Solution:

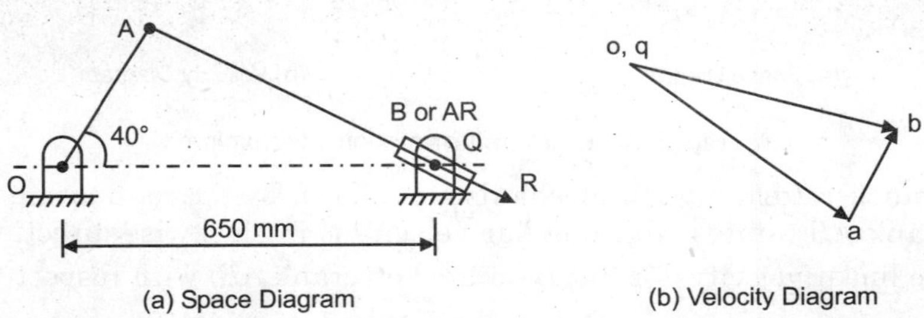

(a) Space Diagram

(b) Velocity Diagram

Fig. 2.9

Select a suitable scale and draw space diagram with the following data.

$$OA = 300 \text{ mm} = 0.3 \text{ m}$$
$$OB = 650 \text{ mm} = 0.65 \text{ m}$$
$$\angle AOB = 40°$$

Measure length of link AB = 460 mm = 0.46 m

Angular velocity of link OA = 20 rad/s

$v_{A/O} = v_A = \omega \times OA = 20 \times 0.3 = 6$ m/s.

Construction of Velocity Diagram

1. Select a suitable scale and draw $\overrightarrow{oa}$ from any point o perpendicular to OA.
2. From point a draw $\overrightarrow{ab}$ perpendicular to AB.
3. From point o or q draw $\overrightarrow{ob}$ parallel to AB.
4. The vectors $\overrightarrow{ab}$ and $\overrightarrow{ob}$ intersect at point b.
5. Measure vector $\overrightarrow{ab}$ = 2.55

$$\therefore \omega_{AR} = \omega_{AB} = \frac{\overrightarrow{ab}}{AB} = \frac{2.55}{0.46} = 5.54 \text{ rad/s. anticlockwise.}$$

2.7. VELOCITIES IN QUICK RETURN MOTION MECHANISM

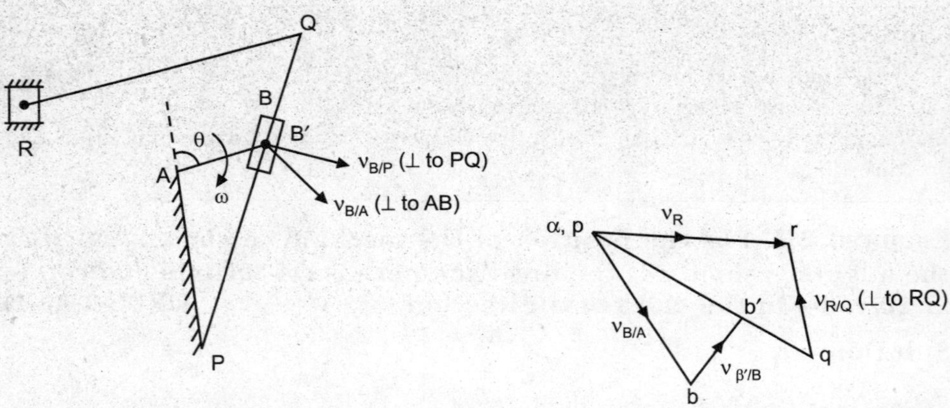

Fig 2.10. Quick return motion mechanism

A quick return motion mechanism is shown by space diagram 2.10(*a*). The crank AB rotates with angular velocity ω in clockwise direction.

The linear velocity of point B (located on crank AB) with respect to A will be

$$v_{B/A} = \omega \times AB \text{ and perpendicular to link } AB.$$

The point A and P are fixed.

$$v_{P/A} = 0$$

The tool box R is moving horizontally. Point B' is on the link PQ. Draw the space diagram to a suitable scale.

Construction of Velocity Diagram

1. Select a suitable scale and from point a anywhere draw victor $\overrightarrow{ab}$ perpendicular to AB to represent $v_{B/A}$.
2. From point p (or a) draw $\overrightarrow{pb'}$ perpendicular to PB' (or PQ) to respect $v'_{B'/P}$.
3. From point b draw $\overrightarrow{bb'}$ parallel to PQ to respect $v_{B'/B}$. The vectors $\overrightarrow{pb'}$ and $\overrightarrow{bb'}$ intersect at b'.
4. Extend vector pb' to q to represent $v_{Q/P}$ or v_Q in the following ratio.

$$\frac{\overrightarrow{pb'}}{\overrightarrow{pq}} = \frac{PB'}{PQ}$$

$$\therefore \quad \overrightarrow{pq} = \overrightarrow{pb'} \times \frac{PQ}{PB'}$$

5. Draw $\overrightarrow{ar}$ in the horizontal direction to represent v_R.
6. From point q draw $\overrightarrow{qr}$ perpendicular to QR. The vectors $\overrightarrow{ar}$ and $\overrightarrow{qr}$ interest in point r.
7. Measure the following from velocity diagram.

$$\overrightarrow{ar} = v_R$$

$$\overrightarrow{pq} = v_{Q/P}$$

The angular velocity of link PQ

$$\omega_{PQ} = \frac{v_{Q/P}}{PQ}.$$

Example 2.3. **The driving crank AB of a Quick Return Motion Mechanism rotates at a uniform speed of 200 rpm. When the crank makes an angle of 60° with the vertical, find the velocity of tool box R. Also calculate the angular velocity of link PQ.**

Solution: Draw the space diagram of Quick Return Motion Mechanism to a suitable scale.

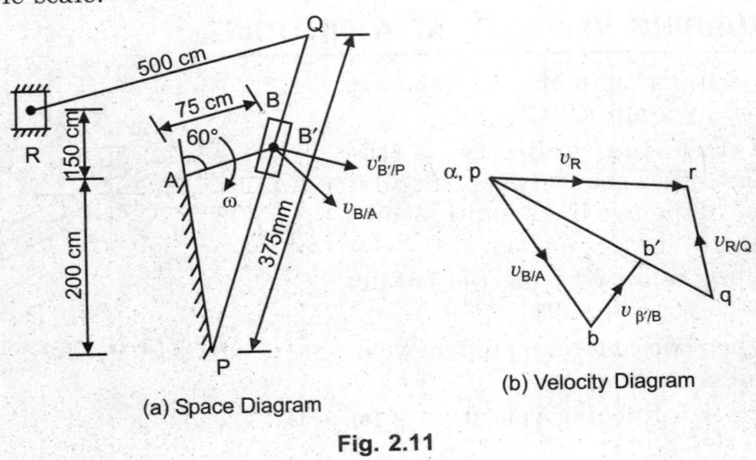

Fig. 2.11

Angular velocity of crank, $\omega_{AB} = \frac{2\pi N}{60} = \frac{2\pi \times 200}{60} = 20.95$ rod/s.

Linear velocity of B w.r.t. A,

$$v_{B/A} = \omega_{BA} \times AB = 20.95 \times 75 = 20.95 \times \frac{75}{1000} = 1.57 \text{ m/s.}$$

Construction of Velocity Diagram

1. Select a suitable scale. From any point a draw $\overrightarrow{ab} = 1.57$ m/s perpendicular to AB.
2. Draw $\overrightarrow{ab'}$ perpendicular to PQ.
3. From points b draw $\overrightarrow{bb'}$ parallel to PQ, Vectors $\overrightarrow{bb'}$ and $\overrightarrow{pb'}$ interest at point b'.
4. From point q draw $\overrightarrow{pr}$ in horizontal direction to represent v_R.
5. Extent vector $\overrightarrow{ab'}$ to q such that

$$\frac{\overrightarrow{pb'}}{\overrightarrow{pq}} = \frac{PB'}{PQ}$$

or

$$\overrightarrow{pq} = \overrightarrow{pb'} \times \frac{PQ}{PB'}.$$

6. From q draw $\overrightarrow{qr}$ perpendicular to QR. The vectors $\overrightarrow{qr}$ and $\overrightarrow{ar}$ intersect at point r.
   The velocity diagram is complete. Measure the following

(*i*) $\overrightarrow{ar} = v_R =$ **1.6 m/s Ans.**

This is the velocity of tool box.

(*ii*) $\overrightarrow{pq} = v_{Q/P} = 1.7$ m/s

Angular velocity of link PQ

$$\omega_{PQ} = \frac{v_{Q/P}}{PQ} = \frac{1.7}{375/1000} = \mathbf{4.533 \text{ rad/s}} \quad \textbf{Ans.}$$

2.8. RUBBING VELOCITY AT A PIN JOINT

The two links in a mechanism are connected by a pin joint.

The **rubbing velocity** is the algebraic difference between angular velocities of the two links multiplied by pin radius.

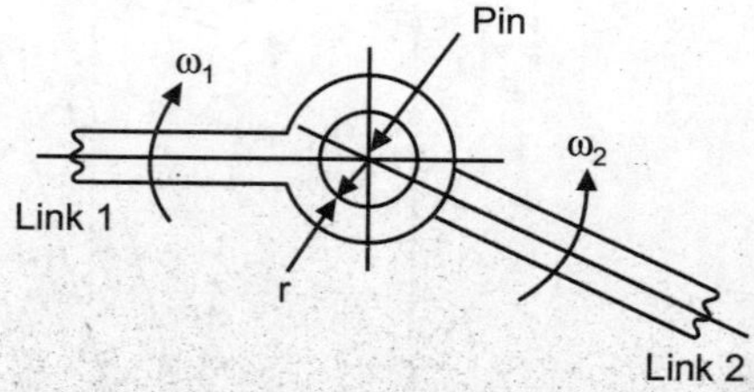

Fig. 2.12 Pin-Joint

Rubbing velocity = $r(\omega_1 \pm \omega_2)$ at pin joint.

1. When two connecting link move in the same direction

$$\text{Rubbing velocity} = r(\omega_1 - \omega_2)$$

2. When two connecting links move is opposite direction

$$\text{Rubbing velocity} = r\ (\omega_1 + \omega_2)\ [\text{m/s}]$$

In the solution of mechanisms, the magnitudes and direction of the angular velocity of the two links of a pin joint are found out. The rubbing velocity can be calculated from the above relations.

2.9. INSTANTANEOUS CENTRE METHOD

The combined motion of rotation and translation can be assumed to be a motion of pure relation about some centre.

This is called **instantaneous centre**. The position of this centric keeps on changing with the position of the rigid body.

The locus of all instantaneous centers is called **centrode**. The position of instantaneous center is located graphically. If instantaneous enter is located, the velocities of all other points on the rigid body or link at any instant can be determined by treating the plane motion as pure rotation about the instantaneous centre.

Example: 2.4. **A link AB is moving in a vertical plane. At a certain moment when link is inclined at 30° with the horizontal, the point *A* is moving horizontally at 4 m/s, while point *B* is moving vertically upwards. Find the velocity of B.**

Solution : Draw perpendiculars to the direction motions at *A* and *B* intersecting at point *I*. Then *I* is the instantaneous centre.

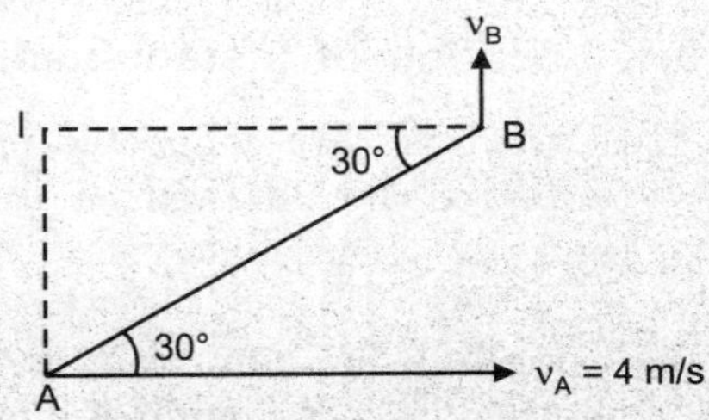

Fig. 2.13 Instantaneous centre

$$\omega = \frac{v_A}{AI} = \frac{v_B}{BI}$$

$$\therefore \quad \frac{v_A}{v_B} = \frac{AI}{BI}$$

In ΔAIB,
$$\tan 30° = \frac{AI}{BI}$$

$$\therefore \quad \frac{v_A}{v_B} = \tan 30°$$

$$\therefore \quad v_B = \frac{v_A}{\tan 30°} = \frac{4}{0.577} = 6.928 \text{ m/s.}$$

Notes:

1. If v_A is known in magnitude and direction and v_B is known in direction, the magnitude of v_B can be determined.
2. The velocity of any point *C* on the link *AB* can be determined in magnitude and direction.
3. The magnitudes of velocities of the point lying on the rigid link is inversely proportional to the distance of the point from the instantaneous centre.
4. The direction of velocity is perpendicular to the line jointing the point to the instantaneous centre.

2.10. TYPES AND LOCATION OF INSTANTANEOUS CENTERS

There are three types of instantaneous centers.

1. Fixed instantaneous center.
2. Permanent instantaneous center.
3. Neither fixed nor permanent instantaneous center.

The number of instantaneous centers is a mechanism is the number of possible combination of two links. If there are n links in a mechanism by taking two links at a true, the number of possible combination can be found out mathematically as.

$$^{n}C_2 = \frac{n}{\underline{|2}\,\underline{|(n-1)}} = \frac{n(n-1)}{2}$$

The number of instantaneous centers,

$$N = \frac{n(n-1)}{2}$$

where n = number of links

For a four-bar mechanism,

$$N = \frac{4(4-1)}{2} = 6.$$

2.10.1. Location of Instantaneous Centers

1. If two links are connected by a pin, the center of the pin will be instantaneous center of permanent type.

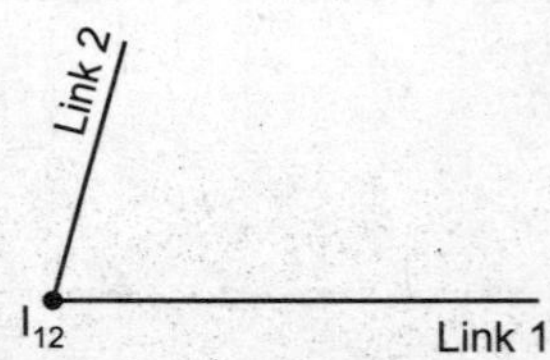

Fig. 2.14 Permanent Contact

2. If one link has pure rolling contact with another link, he point of contact will be instantaneous center.

$$\frac{v_P}{v_Q} = \frac{I_{12}P}{I_{12}Q}$$

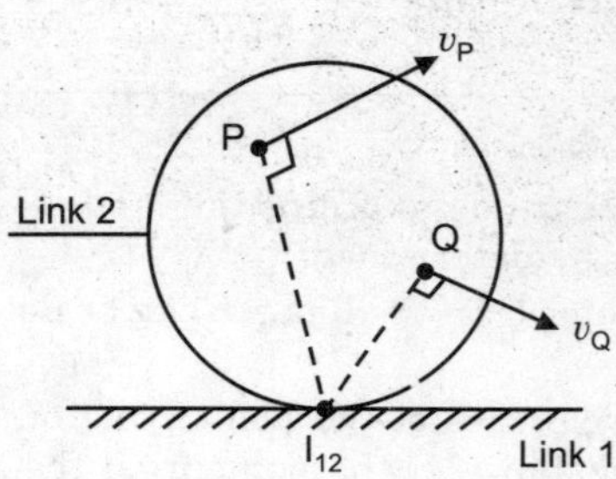

Fig. 2.15. Rolling contact

3. If one link slides over the other link, the instantaneous centre will be on the centre of curvature of curved surface.

$$\frac{v_p}{v_Q} = \frac{I_{12}P}{I_{12}Q}$$

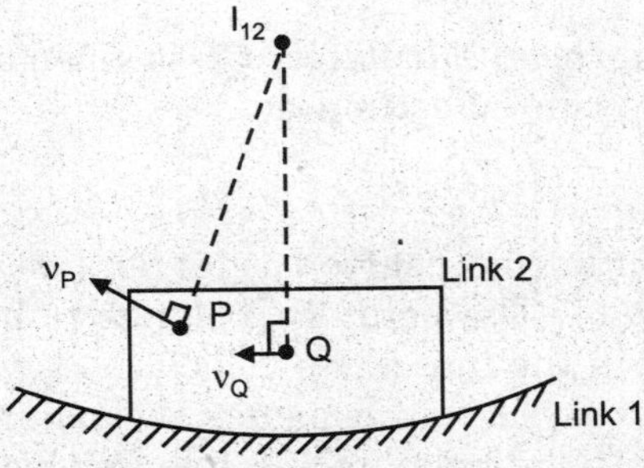

Fig. 2.16 Sliding Contact

2.11. KENNEDY'S THEOREM

If three bodies move relative to each other, they have three instantaneous centres which lie on a straight line. This is also called three centres in line theorem.

A mechanism with three links having relative plane motion is down in Fig. 2.17.

The number of links, n = 3. The number of instantaneous centers

$$N = \frac{n(n-1)}{2} = \frac{3(3-1)}{2} = 3.$$

I_{12} is located at the centre of pin joint of link 1 and 2.

I_{13} is located at the center of pin joint of links 1 and 3. As per Kennedy's theorem, the third instantaneous center must lie on the line joining I_{12} and I_{13}.

In order to plot instantaneous center of rotation, draw lines normal to velocity v_B and v_C. Therefore extend links AB and CD which are normal to velocities v_B and v_C to point I. Point I is the instantaneous centre for link BC.

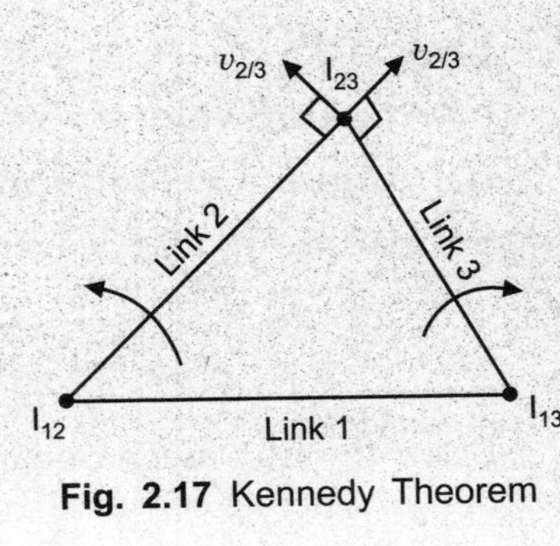

Fig. 2.17 Kennedy Theorem

Now $\quad v_B = \omega_{BC} \times BI \quad$ ...(*i*)

$v_C = \omega_{BC} \times CI \quad$ ...(*ii*)

$$v_B = v_{AB} \times AB = \omega_{BC} \times BI.$$

$$\omega_{BC} = \frac{w_{AB} \times AB}{BI}$$

The value of ω_{AB} and AB are given and value of BI can be measured from Fig. 2.17. Hence ω_{BC} can be found out.

Similarly, $v_C = \omega_{CD} \times CD = \omega_{BC} \times CI$

$$\therefore \quad \omega_{AB} = \frac{\omega_{BC} \times CI}{CD}$$

The value of ω_{BC} has been calculated, CD is given and CI is obtained from Fig 2.17. Value of ω_{CD} can be estimated.

Proof:

Assume that instantaneous centre I_{23} lies outside the line joinign I_{12} and I_{13} as shown in Fig. 2.17. The point I_{23} belongs to both the links 2 and 3. Consider the point I_{23} in the link 2. Its velocity $v_{2/3}$ will be perpendicular to the line joning I_{12} and I_{23}. Again consider the point I_{23} on the link 3. Its velocity $v_{2/3}$ will be perpendicular to the line joining I_{12} and I_{23}.

The velocity of point I_{23} cannot be perpendicular to both lines $I_{12} - I_{23}$ and $I_{13} - I_{23}$. This is possible if point I_{23} lines on the line joining the points I_{12} and I_{13}. hence the three instantaneous centres (I_{12}, I_{13} and I_{23}) must lie on the same straight line. The exact location of I_{23} on line $I_{12} - I_{13}$ depends upon the directions and magnitudes of the three angualr velocities of links 2 and 3 relative to 1.

Example 2.5. ***A* four-bar mechanism has the following dimensions :**

Fixed link $AD = 4$ **m**

Driving link $AB = 1.5$ **m**

Driven link $CD = 2.5$ **m**

Connecting link $BC = 3$ **m**

$\angle BAD = 60°$

Link *AB* rotates at 25 rpm. Determine

(*i*) Angular velocity of link *CD*

(*ii*) Angular velocity of link *BC*.

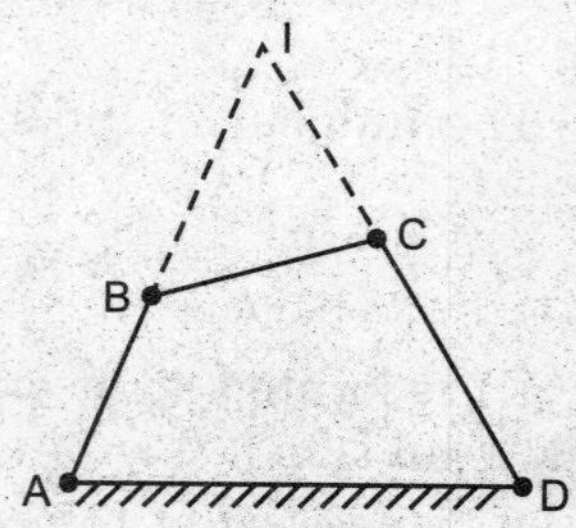

Fig. 2.18

Solution: Angular velocity of link AB

$$\omega_{AB} = \frac{2\pi N}{60} = \frac{2\pi \times 25}{60} = 2.619 \text{ m/s.}$$

Draw the link mechanism to a suitable scale and produce AB and DC to get instantaneous center I. By measurement.

$$CI = 3 \text{ m and } BI = 4.65 \text{ m}$$

$$\therefore \quad \omega_{BC} = \frac{\omega_{AB} \times AB}{BI} = \frac{2.169 \times 1.5}{4.65} = \mathbf{0.845 \text{ rad/s Ans.}}$$

$$\omega_{CD} = \frac{\omega_{BC} \times CF}{CD} = \frac{0.845 \times 3}{4.65} = \mathbf{1.014 \text{ rad/s Ans.}}$$

2.12. VELOCITIES IN FOUR BAR MECHANISM (BY INSTANTANEOUS CENTRE METHOD)

A four-bar mechanism *ABCD* consists of a fixed link *AD*, movable links *AB* and *CD* rotating about points *A* and *D* respectively and a coupler link *BC*. The link *AB* and *CD* rotate whereas link *BC* rotate as well as translate.

Let ω_1 = Angular velocity of link *AB* about A clockwise.

$= \omega_{AB}.$

ω_2 = Angular velocity of link *CD* about *D* clockwise

$= w_{CD}$

ω_{BC} = Angular velocity of link *BC* to be calculated.

v_B = Linear velocity of point *B* perpendicular to *AB*

$= \omega_{AB} \times AB = \omega_1 \times AB.$

v_C = Linear velocity of point *C* perpendicular to *CD*

$= \omega_{CD} \times CD = \omega_2 \times CD.$

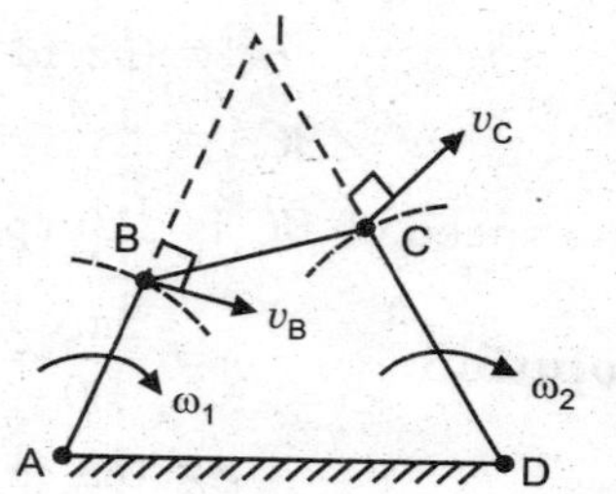

Fig 2.19 Four-Bar Mechanism

2.13. VELOCITIES IN SLIDER CRANK MECHANISM (BY INSTANTANEOUS CENTRE METHOD)

A slider crank mechanism *ABC* has *BC* as crank, *AB* as connecting rod and *A* as piston.

The crank *BC* rotates with angular velocity ω in clockwise direction. The piston *A* reciprocates horizontally. The connecting rod *AB* has combined motion of translation and rotation.

$$\omega = \frac{2\pi N}{60}(\text{rad/s})$$

Linear velocity of *B*

$$v_B = \omega \times BC = \omega_r \perp \text{ to } CB.$$

Linear velocity of *A* is v_A in the horizontal direction.

The position of instantaneous centre *I* can be obtained by drawing perpendiculars to velocity v_A and v_B as shown.

If angular velocity of connecting rod *AB* about $I = \omega_{AB}$

$$v_A = \omega_{AB} \times AI$$

$$v_B = \omega_{AB} \times BI$$

Now $$v_A \doteq \omega_r = \omega_{AB} \times AI$$

∴ $$\omega_{AB} = \frac{\omega r}{AI}$$

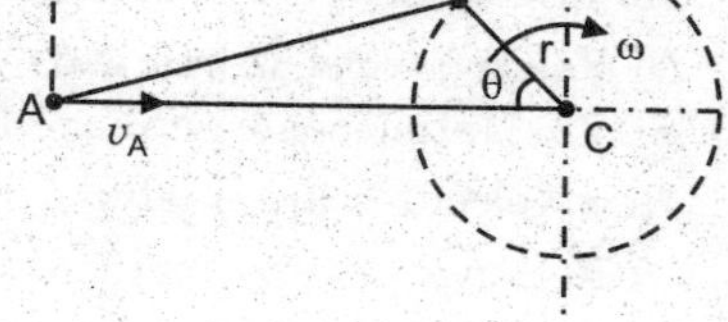

Fig 2.20 Slider Crank Mechanism

The values of ω and r are given, AI can be obtained from Fig. 2.20

$$v_B = \omega_{AB} \times BI = \frac{\omega r}{AI}$$

The values of AI and BI can be measured from Fig 2.20.

Example 2.6. A slider crank mechanism ABC has the following dimension. Find the velocity of point A.

$$AB = 1.0 \text{ m}$$

$$BC = 0.2 \text{ m}$$

$$\angle BCA = 45°$$

The speed of BC is 210 rpm.

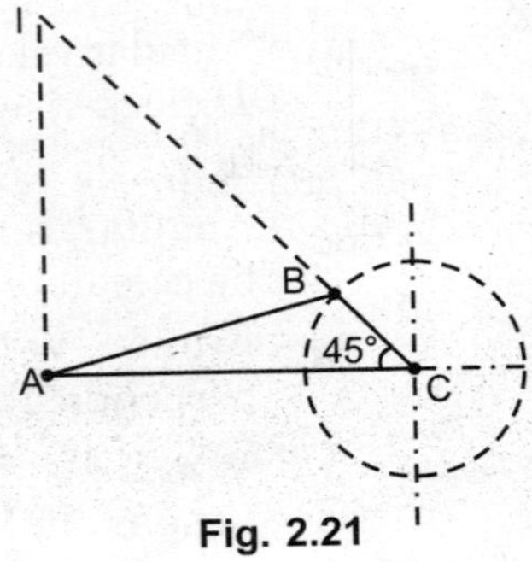

Fig. 2.21

Solution:

$$\omega = \frac{2\pi N}{60} = \frac{2\pi \times 210}{60} = 22 \text{ rad/s.}$$

$$r = 0.2 \text{ m} = 20 \text{ cm}$$

$$AB = 1 \text{ m}$$

$$\theta = 45°$$

Draw slider crank mechanism to a suitable scale. Produce line CB to I and draw a perpendicular AI on line AC. The intersecting point I is instantaneous center.

Measure $AI = 1.15$ m

$BI = 1.41$ m

$$v_A = \frac{\omega r \times AI}{BI} = \frac{22 \times 0.2 \times 1.15}{1.41} = \mathbf{3.58 \text{ m/s}} \quad \textbf{Ans.}$$

QUESTION BANK

1. What is the importance of velocity analysis in mechanisms? Explain various methods.
2. Prove that general plane motion consists of translation and rotation.
3. What is relative velocity method? Define relative velocity of a rigid link.
4. How velocities in a four-bar mechanism are analyzed by
   (i) Relative velocity method.
   (ii) Instantaneous center method.
5. How velocities are analyzed is a slider crank mechanism by
   (i) Relative velocity method
   (ii) Instantaneous center method.
6. State and prove the Kennedy's theorem as applicable to instantaneous centre of rotation of three bodies. **[UPTU 2010–11]**
7. What is velocity of rubbing? How is it found? **[UPTU 2010–2011]**
8. How are velocities in quick return motion mechanism analyzed?
9. Explain types and location of instantaneous centers.

NUMERICAL PROBLEMS

1 In a slider crank mechanism, the length of crank *OB* is 125 mm and that of connecting rod *AB* as 500 mm. The centre of gravity *G* of connecting rod is 275 mm from slider *A*. The crank speed is 600 rpm clockwise. When the crank has made 45° from the inner dead centre position, determine.

(*i*) Velocity of slider *A*.

(*ii*) Velocity of point *G*.

(*iii*) Angular velocity of connecting rod *AB*.

[**Ans.** (*a*) 6.45m/s ; (*b*) 6.75m/s ; (*c*) 10.8 rad/s]

2. Locate the instantaneous centre for a four bar link mechanism shown. *AD* = 125 mm, *AB* = 62.5 mm, *BC* = *CD* = 75 mm. If the link *AB* rotates at a uniform speed of 10 rpm in the clockwise direction find the angular velocity of the links *BC* and *CD*.

[**Ans.** 0.63 rad/s; 0.65 rad/s]

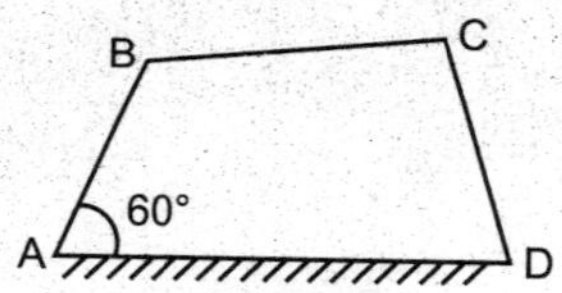

Fig. 2.22

3. In a four-bar mechanism the crank *AB* rotates at 36 rad/s. The length of the links are *AB* = 200 mm; *BC* = 400 mm; *CD* = 450 mm and *AD* = 600 mm. *AD* is the fixed link, At the instant when *AB* is at right angle to *AD*, determine the velocity of.

(*a*) The mid-point of link *BC*.

(*b*) A point on link *CD*, 100 mm from the pin connecting the link *CD* and *AD*.

[**Ans.** 6.55 m/s, 1.45 m/s]

❑❑❑

UNIT–II

3. ACCELERATION IN MECHANISMS

4. MECHANISMS WITH LOWER PAIRS

Chapter 3

ACCELERATION IN MECHANISMS

3.1. INTRODUCTION

Kinematic analysis of mechanisms consists of determination of displacements, velocities and accelerations of links or points in a mechanism for a given input motion. This can also be understood as to find the required input motion to produce a specified motion of another link of the mechanism. Therefore, kinematic analysis is the relationship between the motion of various links of a mechanism.

Acceleration is the rate of change of velocity. The acceleration can be easily determined from known velocities of links. The inertia forces and external forces can be determined after knowing the acceleration as product of mass and acceleration. The forces are required for complete stress analysis of the mechanism and its component.

The acceleration of any point in a mechanism can be determined by analytical methods, as well as graphical method known as acceleration diagram method. The analytical methods sometimes become complex due to unreliable determination of displacements, therefore, acceleration diagram method is more widely used.

3.2. ACCELERATION OF A POINT ON A LINK

When a rigid link rotates about a fixed axis, every point on the link move along concentric circles. When a point on the link moves from position A to B, the angular displacement is measured in radians.

The rate of change of displacement with time is called **angular velocity (ω)**.

$$\omega = \frac{d\theta}{dt}$$

The rate of change of angular velocity with time is called **angular acceleration (α)**.

$$\alpha = \frac{d\omega}{dt} = \frac{d^2\theta}{dt^2}$$

Also $$\alpha = \frac{d\omega}{dt} = \frac{d\omega}{d\theta} \times \frac{d\theta}{dt}$$

$$= \omega \frac{d\omega}{d\theta}$$

Now $$s = r\theta$$

The tangential velocity of the point is called the **linear velocity** (v).

$$v = \frac{ds}{dt} = r\frac{d\theta}{dt}$$

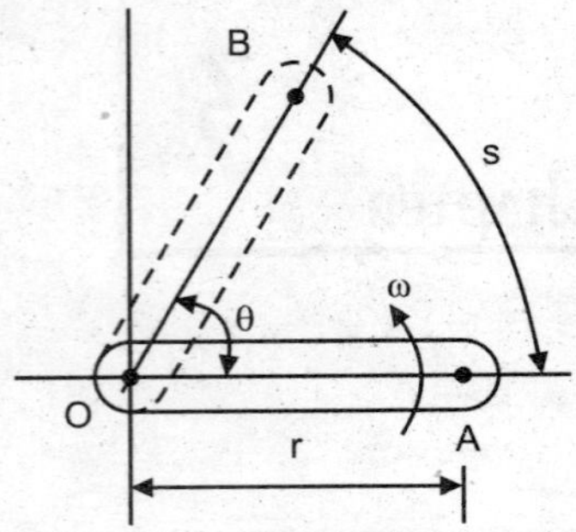

Fig. 3.1 Rotation of a link

The tangential acceleration of the point is called **linear acceleration** (α_t)

$$\alpha_t = \frac{dv}{dt}$$

$$= r\frac{d^2\theta}{dt^2}$$

The radial acceleration

$$\alpha_r = \frac{v^2}{r} = r\omega^2 .$$

Linear velocity of a point

$$v = \omega r \text{ (m/s)}$$

where ω = angular velocity of the link (rad/s)

r = radius of circular path (m)

Linear acceleration

$$a = \frac{dv}{dt} = \frac{d}{dt}(\omega r)$$

$$= r\frac{d\omega}{dt} = r\alpha$$

or $$\alpha = \frac{a}{r}.$$

3.3. ACCELERATION DIAGRAM

Assume a rigid link AB is rotating anticlockwise with angular velocity ω and angular acceleration α. The point A is fixed, point B is rotating about A. The velocity of B is changing in magnitude and direction. Therefore acceleration of point B will have two components.

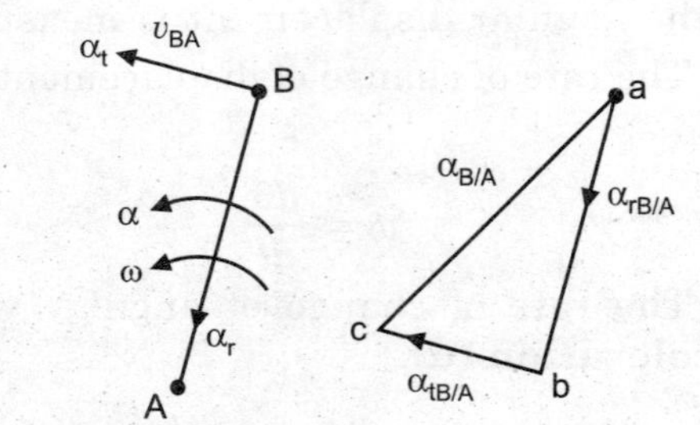

(a) Space Diagram (b) Acceleration Diagram

Fig. 3.2. Acceleration diagram of a link

3.4. ACCELERATION DIAGRAM FOR SLIDER CRANK MECHANISM

Crank AB rotates clockwise with uniform angular velocity ω about fixed point A and makes an angle θ with inner dead centre, CB is the connecting rod.

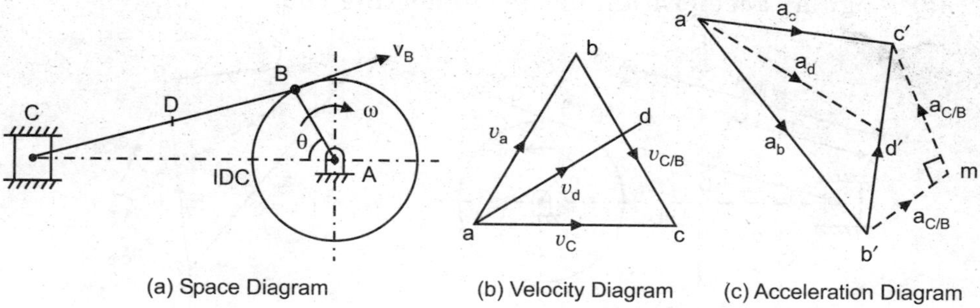

Fig. 3.3. Acceleration diagram of slider crank mechanism

1. Select a suitable scale and take point a' to represent fixed point A. From a' draw $a'b'$ parallel to BA to represent radial component of acceleration of B w.r.t. A.

$$a_b = \frac{v_B^2}{AB}$$

2. From b' draw $\overrightarrow{b'm}$ parallel to CB to represent radial component of C w.r.t.

$$\overrightarrow{b'm} = a_{C/B} = \frac{v_{C/B}^2}{BC}$$

3. From point m draw $\overrightarrow{mc'}$ perpendicular to BC to represent tangential component of acceleration of C w.r.t. B.

$$\overrightarrow{mc'} = a_{C/B}$$

4. From point a' draw $\overrightarrow{a'c'}$ parallel to CA to represent acceleration of C w.r.t. A. The vectors $\overrightarrow{mc'}$ and $\overrightarrow{a'c'}$ intersect at point c'.
5. Triangle $a'b'c'$ is the acceleration diagram of slider crank mechanism.
6. Total acceleration of point D on the connecting rod

$$\frac{\overrightarrow{a'd'}}{\overrightarrow{b'c'}} = \frac{BD}{BC}$$

$$\overrightarrow{b'd'} = \frac{BD}{BC} \times \overrightarrow{b'c'}$$

7. Angular acceleration of connecting rod CB.

$$a_{C/B} = a_{C/B} \times CB$$

$$a_{C/B} = \frac{\alpha_{C/B}}{CB} = \frac{\overrightarrow{a'c'}}{CB}$$

Example 3.1. **The crank of a slider crank mechanism of length 15 cm rotates at 300 rpm clockwise. The connecting rod is 60 cm. When the crank makes an angle 45° from the inner dead centre position, calculate**

(*a*) Acceleration of the midpoint of the connecting rod.

(*b*) Angular acceleration of the connecting rod.

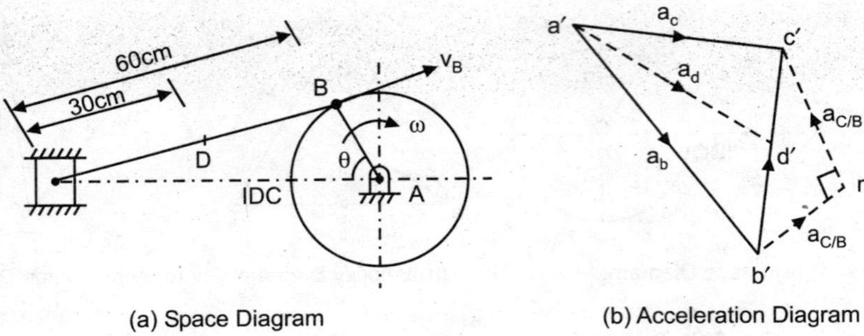

Fig. 3.4.

Solution :

$$AB = 15 \text{ cm}$$
$$BC = 60 \text{ cm}$$
$$CD = 30 \text{ cm}$$
$$N = 300 \text{ rpm}$$
$$\omega = \frac{2\pi N}{60} = \frac{2\pi \times 300}{60} = 31.42 \text{ rad/s}$$
$$\theta = 45°$$
$$v_{B/A} = v_B = \omega \times AB = 31.42 \times 15 = 471.5 \text{ cm/s.}$$

1. Select a suitable scale and draw space diagram.
2. Choose a suitable scale and take a point a' to represent fixed point A. Draw $\overrightarrow{a'b'}$ parallel to BA to represent $a_{B/A}$. Cut $\overrightarrow{a'b'} = \frac{v_2^B}{AB} = \frac{(471.5)^2}{15}$ $= 14808 \text{ cm/s}^2$.
3. From point b' draw $\overrightarrow{a'm}$ parallel CB to represent radial component of acceleration of C w.r.t. B.
4. From point a' draw $\overrightarrow{a'c'}$ parallel to CA to represent acceleration of C w.r.t. A.
5. From point b' cut $\overrightarrow{b'm} = \frac{v_{C/B}^2}{CB} = \frac{340^2}{60} = 1926.7 \text{ cm/s}^2$.
6. From point m draw $\overrightarrow{ma'}$ perpendicular to $\overrightarrow{b'm}$. The vectors $\overrightarrow{ma'}$ and $\overrightarrow{a'c'}$ cut at point c'.
7. $\Delta\, a'b'c'$ is the acceleration diagram of slider crank mechanism.
8. Take midpoint of $\overrightarrow{c'b}$ as d'. Join a' to d'.
9. Measure $\overrightarrow{a'd'} = 11700 \text{ cm/s}^2$

$$\overrightarrow{mc'} = 10300 \text{ cm/s}^2.$$

10. Acceleration of point D, $a_D = \overrightarrow{a'd'} = 11700 \text{ cm/s}^2$
11. Angular acceleration of connecting rod

$$\alpha_{C/B} = \frac{a_{C/B}}{CB} = \frac{10300}{60} = \mathbf{171.67 \text{ rad/s}^2 \text{ Ans.}}$$

Angular Acceleration

$$a_{B/A} = 0 \text{ as } AB \text{ has uniform velocity}$$

1. Total accelration of B w.r.t. A,
$$aB/A = a_B = aB/A \text{ (radial)}$$
2. Select a suitable scale and plot point a' anywhere. From point a' draw $\overrightarrow{a'b'}$ parallel to AB to represent radial component of acceleration of B w.r.t. and cut equal to magnitude of $a\,B/A$.
3. From piont b' draw $b'm$ parallel to BC to represent radial component of acceleration of C w.r.t. B and cut equal to aC/B.
4. From point d' draw $\overrightarrow{d'n}$ parallel to CD to represent radial component of C w.r.t. D and cut equal to $a_{C/B}$. Which is equal to $\overrightarrow{mc'}$.
5. From point m draw $\overrightarrow{mc'}$ perpendicular to BC to represent tangential component of acceleration of C w.r.t. B line $\overrightarrow{mc'}$ is perpendicular to line $\overrightarrow{mb'}$.
6. From point n draw $\overrightarrow{nc'}$ perpendicular to CD to represent the tangential component of acceleration of C w.r.t. D.
7. $\Delta\, a'b'c'$ is the acceleration diagram of four-bar mechanism.
8. Measure the following

$$\overrightarrow{mc'} = a_{C/B}$$

$$\overrightarrow{nc'} = a_{C/D}$$

9.

$$\alpha_{B/C} = \frac{a_{C/B}}{CB}$$

$$\alpha_{C/D} = \frac{a_{C/D}}{CD}$$

10. Locate point e' on $\overrightarrow{b'c'}$ so that

$$\frac{\overrightarrow{b'e'}}{\overrightarrow{b'c'}} = \frac{BE}{BC}$$

Join $\overrightarrow{a'e'}$ and measure it

$$\therefore \quad a_E = \overrightarrow{a'e'}$$

3.5. ACCELERATION DIAGRAM OF FOUR-BAR MECHANISM

A four bar-mechanism *ABCD* has *AD* as fixed. The crank *AB* rotates at *N* rpm and drives the link *CD* with the help of connecting rod *BC*. Draw the acceleration diagram and find acceleration of point *E* in link BC.

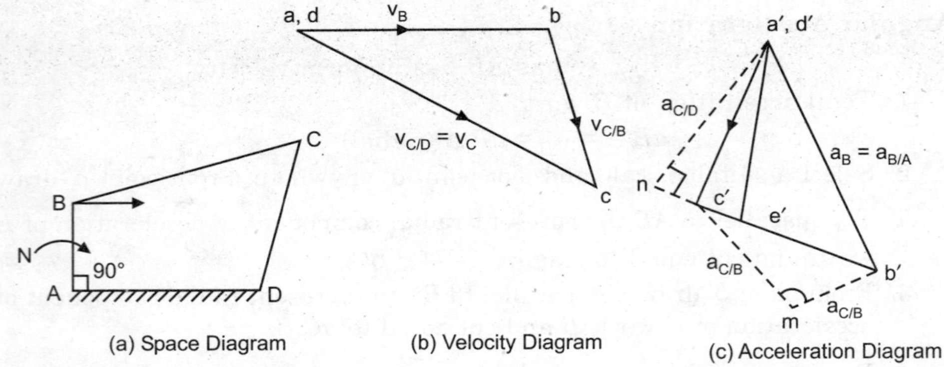

Fig. 3.5. Acceleration Diagram for four-bar mechanism

Angular velocity of link *AB*,

$$\omega_{BA} = \frac{2\pi N}{60}$$

Velocity of *B* w.r.t. *A*,

$$v_{B/A} = v_B = \omega_{BA} \times BA$$

1. Space diagram

Select a suitable scale and draw the space diagram of the mechanism as per the given dimensions.

2. Velocity diagrams

(*i*) Select asuitable scale and plot point a (and *d*).

(*ii*) Draw vector $\overrightarrow{ab}$ to represent $v_{B/A}$ or v_B.

(*iii*) From point *b* draw $\overrightarrow{bc}$ perpendicular to *BC*.

(*iv*) From point *d* draw $\overrightarrow{dc}$ perpendicular to *DC* to intersect $\overrightarrow{bc}$ in *c*.

(*v*) Δabc is the velocity diagram.

(*vi*) Measure the following

$$\overrightarrow{bc} = v_{C/D} = v_C$$
$$\overrightarrow{ab} = v_{B/A} = v_B$$

3. Acceleration diagram

***(i)* Calculation:** Calculate radial component of acceleration of every link.

$$a_{B/A} = \frac{v_{B/A}^2}{BA}$$

$$a_{C/B} = \frac{v_{C/B}^2}{CB}$$

$$a_{C/D} = \frac{v_{C/D}^2}{CD}$$

1. Radial component (α_r) due to angular velocity (ω) acting parallel to BA from B to A.

$$\alpha_r = \frac{v^2}{r} = \omega^2 r$$

$$\therefore \qquad a_{r_{B/A}} = \frac{v_{B/A}^2}{BA} = \omega^2 \times BA \qquad [BA = r]$$

2. Tangential component (α_t) due to angular acceleration (α) acting perpendicular to AB.

$$\alpha_t = \alpha \times BA$$

$$\alpha_{t_{B/A}} = \alpha \times BA.$$

3. The total (resultant) acceleration of B w.r.t. A is the vector sum of the two components

$$\alpha_{B/A} = \vec{\alpha}_{r_{B/A}} + \vec{\alpha}_{t_{B/A}}$$

$$= \frac{v_{B/A}^2}{BA} + \alpha \times BA.$$

Note: If B is rotating with uniform angular velocity (ω) and angular acceleration (α) is zero.

$$\alpha_{t_{B/A}} = 0$$

$$\alpha_{B/A} = \frac{v_{B/A}^2}{BA}.$$

Construction of Acceleration Diagram

1. Select a suitable scale and take point (fixed) a anywhere. From a draw vector $\overrightarrow{ab}$ parallel to BA to represent the radial component of acceleration of point B w.r.t. A., *i.e.*, $\alpha_{rB/A}$.
2. From point b draw vector $\overrightarrow{bc}$ perpandicular to AB to represent tangantial component of acceleration of B w.r.t. A *i.e.*, $\alpha_{t_{B/A}}$.
3. Join a to c. Then vector $\vec{ac}$ represents the total acceleration of B w.r.t. A in magnitude and direction, *i.e.* $\alpha_{B/A}$.

Note: Use small letters in the acceleration diagram and capital letter in space diagram.

3.6. CORIOLIS COMPONENT OF ACCELERATION

In quick return motion mechanism, a point on one link is sliding along another rotating link, Coriolis component of acceleration is created and must be calculated.

A link AR is rotating about a fixed point A with angular velocity ω and angular acceleration α. There is a slider on the link and P is a point on the slider and point Q on the link AP. Assume the following.

v = Linear velocity of the slider on the link

a = Linear acceleration of the slider on the link.

r = Radial distance of point P on the slider from fixed point A.

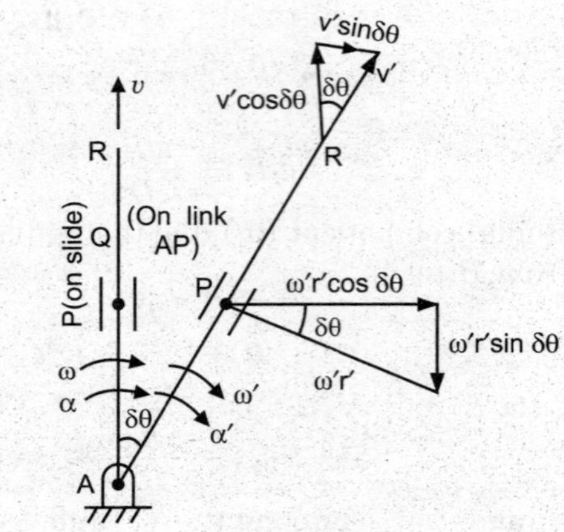

Fig. 3.6. Corliolis Acceleration component

In an interval of time δt (very short), $\delta\theta$ is the angular displacement of the link and δr is the outward radial displacement of the slider.

The angular velocity of link $\omega' = \omega + a\delta r$

The linear velocity of the slider on the link, $v' = v + a\delta t$

Radial distance of the slider, $r' = r + \delta r$.

3.6.1. Acceleration of *P* Parallel to *AR*

Initial velocity of P along $AR = v = v_{p/q}$

Final velocity of P along $AR = v' \cos \delta\theta - \omega' r' \sin \delta\theta$

Change of velocity along $AR = (v' \cos \delta\theta - \omega' r' \delta\theta) - v$

$$\text{Acceleration of } P \text{ along } AR = \frac{(v + a\delta t)\cos\delta\theta - (\omega + \alpha\delta t)(r + \delta r)\sin\delta\theta - a}{\delta t}$$

When $\delta t \to 0$, $\cos \delta\theta \approx 1$ and $\sin \delta\theta \approx \delta\theta$.

$$\text{Acceleration of } P \text{ along } AR = a - \omega r \frac{d\theta}{dt}$$

$$= a - \omega r \omega = a - \omega^2 r$$

Accelaration of P along AR = Acceleration of slider – radial acceleration in outward radial direction.

3.6.2. Acceleration of *P* Perpendicular to *AR*

Intial velocity of $P \perp AR = \omega r$

Final velocity of $P \perp AR = v' \sin \delta\theta + \omega' r' \cos \delta\theta$

Change of velocity of $P \perp AR = (v' \sin \delta\theta + \omega' r' \cos \delta\theta) - \omega r$

$$\text{Acceleration of } P \perp AR = \frac{(v + a\delta t)\cos\delta\theta - (\omega + \alpha\delta t)(r + \delta r)\cos\delta\theta - \omega r}{\delta t}$$

When $\delta r \to 0$, $\cos \delta\theta \to 1$ and $\sin \delta\theta \to 1$.

$$\text{Acceleration of } P \perp AR = v\frac{d\theta}{dt} + \omega\frac{dr}{dt} + r\alpha$$

$$= v\omega + \omega v + r\alpha = 2\omega v + r\alpha$$

$$= 2\omega v + \text{Tangential acceleration.}$$

3.6.3. Coriolis Acceleration Component

The component $2\omega v$ is called Coriolis acceleration component. The Coriolis acceleration component is positive if both ω and v are positive or both are negative.

(*i*) The link *AR* should rotate clockwise and slider should move radially outwards.

(*ii*) The link *AR* should rotate anticlockwise and slider should move radially inward.

Rule: The direction of Coriolis acceleration component can be obtained by rotating the radial vector $\vec{v}$ through 90° to the direction of rotation of the link.

Acceleration of *P* = Acceleration of *P* parallel to AR + Acceleration of *P* parpendicular to *AR*

$$a_P = (a - \omega^2 r) + (2\omega v + r\alpha)$$
$$= a + (r\alpha - \omega^2 r) + 2\omega v$$
$$= a_{P/Q} + a_{Q/A} + a_{CR}$$

where *Q* is a point on the link *AR* immediately below the point *P* on the slider.

$$a_{P/A} = a_{P/Q} + a_{Q/A}$$

where $\quad a_{P/a} = a'_{P/Q} + a_{CR}$

3.7. CRANK AND SLOTTED LEVER MECHANISM

The crank *OP* of a crank and slotted lever mechanism rotates uniformly clockwise with angular velocity ω.

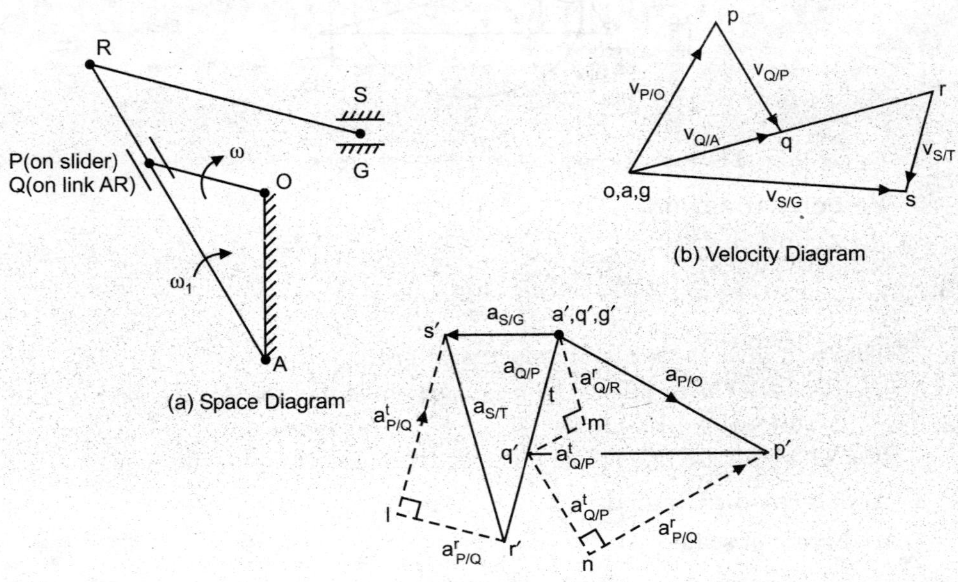

(a) Space Diagram

(b) Velocity Diagram

(c) Acceleration Diagram

Fig. 3.7. Crank and slotted lever mechanism

1. Select a suitable scale and draw the space diagram.
2. Draw the velocity diagram.
3. Calculate the radial components of acceleration of all the links as $a^r = \dfrac{v^2}{\text{Link length}}$.
4. Draw the acceleration diagram. All radial components of acceleration will be parallel to the corresponding links and tangential components of acceleration perpendicular to the direction of corresponding links.
5. Measure the magnitudes of total acceleration of the links.
6. Calculate the angular acceleration as

$$\alpha = \frac{\text{Liner acceleration}}{\text{Length of link}} = \frac{a}{\text{Length}}$$

3.8. KLEIN'S CONSTRUCTION

Prof. Klein proposed a simple graphical construction as velocity and acceleration diagrams. These diagrams are directly made on the space diagram of the mechanism.

3.8.1. Klein's Construction for Slider Crank Mechanism

1. Draw the space diagram to a suitable scale from the known dimensions. *ABC* is the slider crank mechanism.

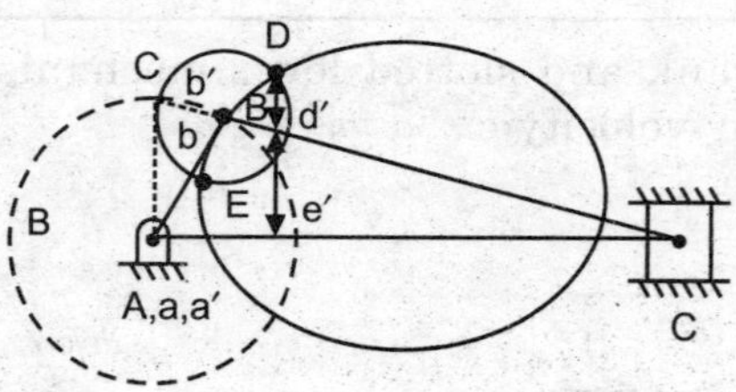

Fig. 3.8. Klein's construction for slider crank mechanism

2. **Velocity Diagram**

   (*i*) Fixing of scale

$$AB = r = v_{B/A}$$

$$\therefore \quad \overrightarrow{ab} = AB = \omega r$$

   This will give the scale of velocity diagram.

   (*ii*) Produce *CB* to draw a link *ac* perpendicular to *AC* intersecting the produced live *CB* in *C*.

   (*iii*) The triangle *abc* is the velocity triangle of the mechanism.

3. **Acceleration Diagram**

   (*i*) Fixing of scale

$$AB = \overrightarrow{a'b'} = \omega^2 r$$

   This gives the scale of acceleration diagram.

   (*ii*) Draw a circle with *b* as centre and *bc* as radius.

(iii) Draw a circle with AC diameter. Points D and E are the points of intersection. Joint DE and extend to meet AB at e'.

(iv) $a'\ b'\ c'\ d'\ e'$ is the acceleration diagram of slider crank mechanism.

$$\overline{b'd'} = a^r_{B/A}$$

3.8.2. Klein's Construction for Four Bar Mechanism

$ABCD$ is a four-bar mechanism with link AD as fixed and link AB as driving crank. For the pins at B and C there is a relation

$$a_c = a_b + a_{C/B}$$

Each links AB, BC and CD have angular velocity and angular acceleration at any instant of time.

1. Draw space digram to a scale as per dimensions given

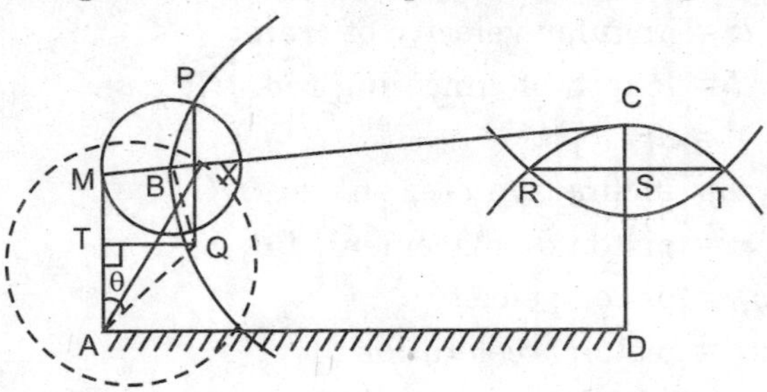

Fig. 3.9. Klein's construction for Four Bar Mechanism

2. **Velocity Diagram**

   Draw AM parallel to DC and produce CB to intersect AM in M,

$$v_B = \omega.AB$$
$$v_C = \omega AM$$
$$v_{C/B} = \omega BM$$
$$\omega_{B/C} = \frac{v_{C/B}}{BC}$$
$$\omega_{C/D} = \frac{v_C}{CD}$$

3. **Acceleration Diagram**

   (i) With B as centre and BM as radius draw a circle.

   (ii) Draw a circle with BC as diameter.

   (iii) Draw PQ as common chord to two circle.

   (iv) The chord PQ intersects BC in X.

   (v) With centre as C and radius AM draw a circle.

   (vi) Draw a circle on CD as diameter. RT is the common chord cutting CD in S.

   (vii) Cut $AT = CS$

   (viii) Draw TQ perpendicular to AM.

   (ix) $a_c = \omega^2 QA$.

   $a_{C/B} = \omega^2 QB$.

3.9. ANALYTICAL METHOD FOR SLIDER CRANK MECHANISM

Figure 3.10 represents a model for analytical solution of velocity and acceleration of a slider crank mechanism

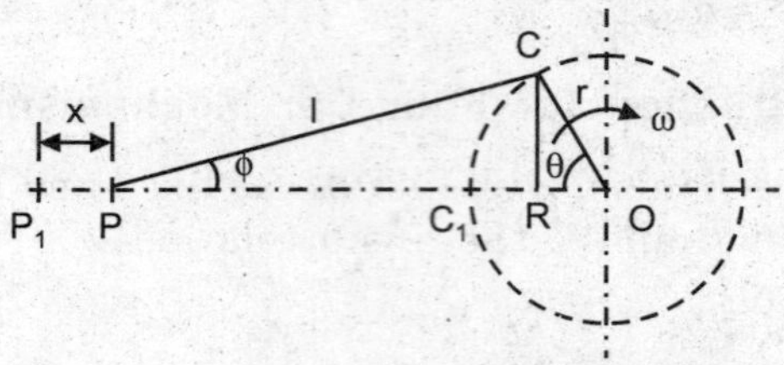

Fig. 3.10. Analyticl model of slider crank mechanism

Let r = length of crank OC

ω = angular velocity of crank

l = length of cnnecting rod, CP

$= nr$

θ = inclination of crank to IDC

x = piston displacement, PP_1

v_P = piston velocity

a_P = piston acceleration

ω_r = angular velocity of connecting rod

α_r = angular acceleration of connecting rod

CR = perpendicular to line of stroke

Piston displacement, $PP_1 = x$

$$\therefore \quad x = P_1O - PO$$

$$= l + r - (l \cos \phi + r \cos \theta)$$

$$= r(1 - \cos \theta) + l(1 - \cos \phi)$$

Now $\quad CR = r \sin \theta = l \sin \phi$

$$\therefore \quad \sin \phi = \frac{r}{l}\sin\theta = \frac{\sin\theta}{n} \quad \text{...(1)}$$

$$\cos \phi = \sqrt{1-\sin^2\phi} = \frac{1}{n}\sqrt{n^2-\sin^2\phi}$$

$$\therefore \quad x = r(1-\cos\theta)+nr\left\{1-\left(\frac{1}{n}\right)\sqrt{n^2-\sin^2\theta}\right\}$$

$$= r\left[1+n-\cos\theta-\sqrt{n^2-\sin^2\theta}\right]$$

$$v_P = \frac{dx}{dt} = \frac{d\theta}{dt}\cdot\frac{dx}{d\theta}$$

$$= \omega r\left\{\sin\theta+\frac{\sin 2\theta}{2\sqrt{n^2-\sin^2\theta}}\right\}$$

$$a_P = \frac{dv_p}{dt} = \frac{d\theta}{dt} \cdot \frac{dv_p}{d\theta}$$

$$= \omega^2 r \left[\cos\theta + \frac{n^2 \cos 2\theta + \sin^4 \theta}{(n^2 - \sin^2 \theta)^{3/2}} \right]$$

$\sin^2\theta$ is very small as compared to n^2

$$\therefore \quad v_p \approx \omega r \left(\sin\theta + \frac{\sin 2\theta}{2n} \right)$$

$$a_p \approx \omega^2 r \left(\cos\theta + \frac{\cos 2\theta}{n} \right)$$

Differenting equation (1)

$$\cos\phi \frac{d\phi}{dt} = \frac{\cos\theta}{n} \cdot \frac{d\theta}{dt}$$

$$\therefore \quad \omega_r = \frac{d\phi}{dt} = \frac{\omega \cos\theta}{n \cos\phi} = \omega \frac{\cos\theta}{\sqrt{n^2 - \sin^2\theta}}$$

and

$$\alpha_r = \frac{d\omega_r}{dt} = \frac{d\phi}{dt} \cdot \frac{d\omega_r}{d\theta}$$

$$= \frac{-\omega^2 (n^2 - 1) \sin\phi}{(n^2 - \sin^2\theta)^{3/2}}$$

Now n^2 is large compared to $\sin^2\theta$ and unity.

$$\omega_r \approx \left(\frac{\omega}{n} \right) \cos\theta$$

$$\alpha_r = \left(-\frac{\omega^2}{n} \right) \sin\theta$$

QUESTION BANK

1. Define acceleration of a point on a rotating link.
2. What is acceleration diagram? How would you draw acceleration diagram for a slider crank mechanism?
3. Detail out a procedure to draw acceleration diagram for a four-bar mechanism.
4. What is Coriolis acceleration component? How is it determined?
5. Explain the procedure to construct Klein's construction to determine velocity and acceleration of a slider crank mechanism. **[UPTU 2010-11]**
6. Explain step-by-step Klein's construction for four-bar mechanism.
7. Detail out an analytical method to determining velocity and acceleration in a slider crank mechanism.

NUMERICAL PROBLEMS

1. For the inverted slider crank mechanism shown below in Fig. 3.11. Find the angular acceleration of link QR. The crank OA is 300 mm long and rotates at 20 rad/s in clockwise direction. **[UPTU 2010-11]**

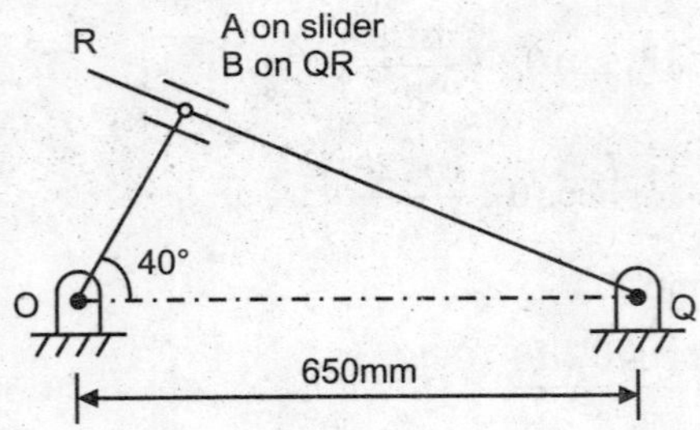

Fig. 3.11.

2. The length of crank and connecting rod of a steam engine are 100 mm and 400 mm respectively. The weight of connecting rod is 50 kg and its centre of mass is 200 mm from the cross-head centre. The radius of gyration about centre of mass is 120 mm. If the engine speed is 300 rpm and crank has turned 45° from the inner dead centre, determine.
   (*a*) Velocity and acceleration of the centre of mass of the connecting rod.
   (*b*) The angular velocity and acceleration of connecting rod.
   (*c*) The kinetic energy of the rod.

{**Ans.** (a) 2.7 m/s, 80 m/s^2 (b) 5.7 rad/s, 173 rad/s (c) 194 N-m}

Chapter 4

MECHANISMS WITH LOWER PAIRS

4.1. INTRODUCTION

Lower pairs usually consist of turning (pin joints) and sliding pairs. Turning pairs are very widely used in mechanisms and machines to produce a wide variety of motions. However, these days, sliders are also used to get linear motions.

Simple mechanisms and their inversions have already been discussed in chapter 1, more common mechanisms with lower pairs have been selected for analysis.

The present chapter will deal with elementary analysis of mechanisms followed by their potential in the design of mechanisms and machines of daily practical use. Systematic design techniques can be developed from these studies so that these mechanisms can be used for accurate control of the processes and the machines required for modern technology.

The following mechanisms developed by different engineers have been selected for analysis.

1. Pantograph.
2. Peaucellier's mechanism for exact straight line motion.
3. Hart mechanism.
4. Scott-Russel mechanism
5. Grass-Hopper mechanism for approximate straight line motion.
6. Watt mechanism
7. Tchebicheff mechanism
8. Hooke's joint
9. Davis steering gear mechanism
10. Achermann steering gear mechanism

4.2. PANTOGRAPH

A pantograph is a four-bar linkage connected by turning pairs. It is an instrument to reproduce to an enlarged or reduced scale plane areas or maps, straight lines or curved paths and as exactly as possible by copying the path described by a gives point.

1. Construction

It consists of a jointed parallelogram $ABCD$ as shown in Fig. 4.1. The four bars are connected by turning pairs (pin joints). The bars BA and BC are extended to O and E respectively, such that

$$\frac{OA}{OB} = \frac{AD}{BE}$$

The point O is made fixed. For all relative positions of the bars, the triangles OAD and OBE are similar and the points O, D and E are in one straight line. The point E traces out the same path as described by point D.

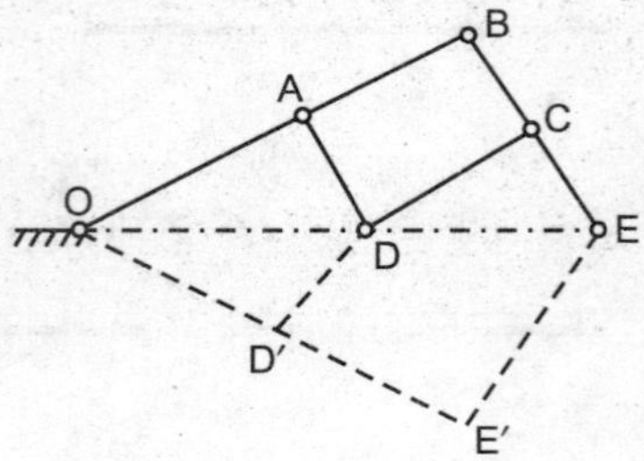

Fig. 4.1. Pantograph

2. Operation

The linkage is moved and ratio is maintained.

$$\frac{OD}{OE} = \frac{AD}{BE}$$

Point O is fixed and the points D and E move to new position D' and E'.

$$\therefore \quad \frac{OD}{OE} = \frac{OD'}{OE'}$$

The straight line DD' is parallel to the straight line EE'. The point O is fixed to the frame of a machine by means of a turning pair and D is attached to a point in the machine which has rectilinear motion relative to frame, then E will trace out a straight line path. Similarly, if E is constrained to move in a straight line, then D will trace out a straight line parallel to the former.

3. Applications

(*i*) A pantograph is mostly used for the reproduction of plane areas and figures such as maps, plans, etc on enlarged or reduced scales.

(*ii*) A pantograph is used as an indicator rig in order to reproduce to a small scale the displacement of piston of a reciprocating steam engine or I.C. engine.

(*iii*) A pantograph is used to guide cutting tools.

(*iv*) A modified pantograph is used to collect power at the top of an electric locomotive.

4.3. STRAIGHT LINE MOTION MECHANISMS

A straight line motion mechanism is a constrained mechanism to permit only relative motion of an oscillatory nature along a straight line. These mechanisms have the following types of pairs.

1. Only turning pairs.
2. One sliding pairs.

These mechanisms can produce the following types of motion

1. Exact straight line motions
   (*i*) Mechanism made up of turning pairs
   Examples: Peaucellier mechanism, Hart's mechanism

(*ii*) Mechanism consisting of one sliding pair.

Examples: Scott-Russel's mechanism

2. Approximate straight line motion mechanisms.

These are modifications of four-bar chain mechanism.

Examples: Watt's mechanism, Scott-Russel's modified mechanism, Tchebicheff's mechanism.

The principle adopted for exact straight line motion or mathematically correct motion is illustrated by Fig. 4.2. O is a point on the circumference of a circle or diameter OP. OA is a chord and B is a point on extended OA such that

$$OA \times OB = \text{Constant}$$

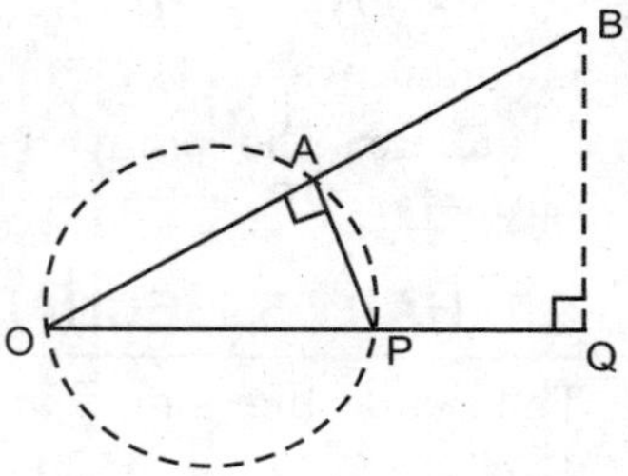

Fig. 4.2. Exact straight line motion mechanism.

Therefore, the locus of point B is a straight line BQ perpendicular to diameter OP.

Proof. Draw BQ perpendicular to line OP extended to Q-Join AP.

The triangles OAP and OBQ are similar

$$\therefore \quad \frac{OA}{OP} = \frac{OQ}{OB}$$

$$\text{or} \quad OP \times OQ = OA \times OB$$

$$\text{or} \quad OQ = \frac{OA \times OB}{OP}$$

$$OP = \text{Constant being diameter of a circle}$$

$$\therefore \quad OA \times OB = \text{Constant}$$

$$\text{and} \quad OQ = \text{Constant}$$

Hence point B moves along straight path BQ which is perpendicular to OP.

4.4. PEAUCELLIER'S EXACT STRAIGHT LINE MOTION MECHANISM

Peaucellier's mechanism is show in Fig. 4.3, which consists of fixed link OO_1, and other straight links O_1A, OC, OD, AD, DB, BC and CA are connected by turning pairs at their intersections. The pin at A is constrained to move along the circumference of a circle with fixed diameter OP, by means of the link O_1A. The following geometrical conditions are satisfied

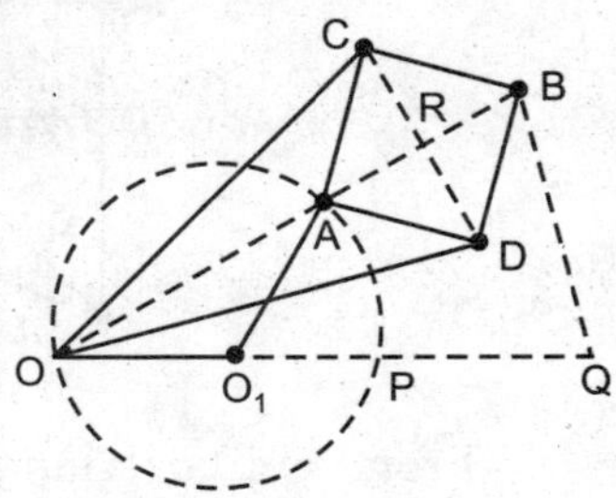

Fig. 4.3. Peaucellier mechanism

$$AC = CB = BD = DA$$

$$\text{and} \quad OC = OD$$

$$OO_1 = O_1A$$

Proof. When link O_1A rotates, the product $OA \times OB$ remains constant.

Join CD to bisect AB at R.

From right angle triangle ORC and BRC,

$$OC^2 = OR^2 + RC^2 \quad ...(1)$$
$$BC^2 = RB^2 + RC^2 \quad ...(2)$$

Subtract (2) from (1)

$$OC^2 - BC^2 = OR^2 - RB^2$$
$$= (OR + RB)\ (OR - RB) = OB \times OA.$$

Now OC and BC have constant lengths, therefore, the product $OB \times OA$ is constant.

Hence the point B traces a straight line path perpendicular to the diameter OP.

4.5. HART'S MECHANISM FOR EXACT STRAIGHT LINE MOTION

The Peaucellier's mechanism consists of eight links whereas Hart's mechanism consists of six links only.

The mechanism consists of fixed link OO_1 and the straight links O_1A, FC, CD, DE and EF are connected by turning pairs at their points of intersection as shown in Fig. 4.4. The following geometrical conditions are satisfied.

$$CD = EF \quad \text{and} \quad FC = DE$$

The points O, A and B divide the links FC, CD and EF in the same ratio.

In triangle FCE, O and B divide FC and EF in the same ratio.

$$\therefore \quad \frac{CO}{CF} = \frac{EB}{EF}$$

$\therefore$ OB is parallel to CE

Similarly in triangle FCD, OA is parallel to FD, OA and OB are respectively parallel to FD and CE in the trapezium $BOCE$.

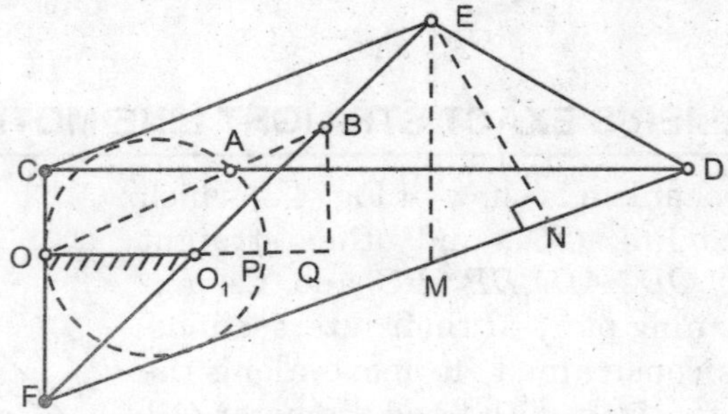

Fig. 4.4. Hart's mechanism

From similar triangles CFE and OFB,

$$\frac{CE}{FC} = \frac{OB}{OF}$$

or,
$$OB = \frac{CE \times OF}{FC} \quad ...(1)$$

From similar triangle FCD and OCA,

$$\frac{FD}{FC} = \frac{OA}{OC}$$

or, $$OA = \frac{FD \times OC}{FC} \quad ...(2)$$

Multiply equations (1) and (2)

$$OA \times OB = \frac{FD \times OC}{FC} \times \frac{CE \times OF}{FC}$$

$$= FD \times CE \times \frac{OC \times OF}{FC^2}$$

Now lengths of OC, OF and FC are fixed $\left(\frac{OC \times OF}{FC^2} = \text{constant}\right)$

$$\therefore \quad OA \times OB = FD \times CE \times \text{constant} \quad ...(3)$$

Draw EM parallel to CF and EN perpendicular to FD.

$$\therefore \quad FD \times CE = FD \times FM \quad (\because CE = FM)$$

$$= (FN + ND)(FN - MN) \quad (\because MN = ND)$$

$$= FN^2 - ND^2$$

$$= (FE^2 - NE^2) - (ED^2 - NE^2)$$ [From right angle triangles FEN and EDN]

$$= FE^2 - ED^2 = \text{constant} \quad ...(4)$$

$\because$ FE and ED are fixed lengths

From equation (3) and (4)

$$OA \times OB = \text{constant.}$$

If the mechanism is pivoted at O as a fixed point and point A is constrained to move in a circle with centre O_1, the point B will trace a straight line perpendicular to the diameter OP extended to Q.

Limitation: The mechanism occupies a large space even though the path of B is short.

4.6. SCOTT RUSSEL'S MECHANISM OF EXACT STRAIGHT LINE MOTION

The Scott-Russel's mechanism is shown in Fig. 4.5 and consists of a fixed member and moving member P of sliding pair.

The straight line PAQ is connected to link OA and link P by turning pairs. The link OA rotates about O.

The mechanism OAP is same as slider-crank mechanism of a reciprocating engine. OA is the crank and PA is the connecting rod. The mechanism copies a straight line motion and not generated.

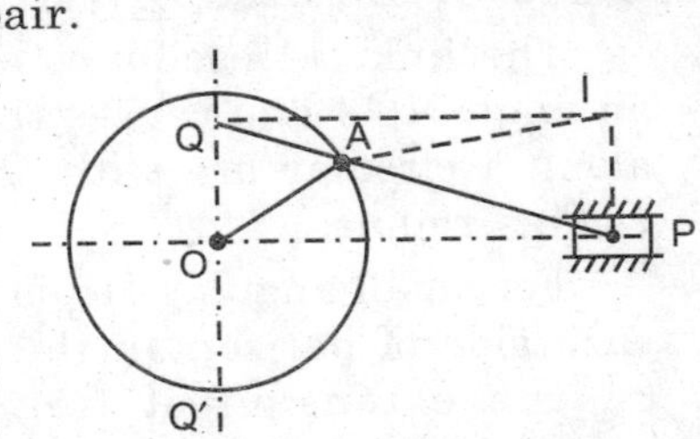

Fig. 4.5. Scott-Russel's mechanism

If A is the middle point of PQ and $OA = AP = AQ$, the instantaneous centre of motion of link PAQ lies at 1 in the OA extended and IP is perpendicular to IQ. Now IQ is perpendicular to OP in the rectangle $OPIQ$, Q moves along the vertical line OQ for all positions of QP. Hence Q traces the straight line QOQ'.

When OA makes one complete revolution, P will oscillate along the line OP through a distance $2 \times OP'$ on each side of O. The point Q will oscillate along QOQ' through the same distance $2 \times OA$ above and below O. Therefore, the locus of Q is a copy of the locus of P.

Limitation: This mechanism does not have much practical use because of high friction loss in the sliding pair as compared to a turning pair. The high wear rate affects the operation and life of the mechanism.

4.6.1. Modified Scott-Russel Mechanism

A modified Scott-Russed mechanism is shown in Fig. 4.6 where AP is not equal to AQ and points P and Q are constrained to move in the horizontal and vertical directions.

It forms an elliptical trammel such that any point A on PQ traces an ellipse with semi-major axis AQ and semi-minor axis AP.

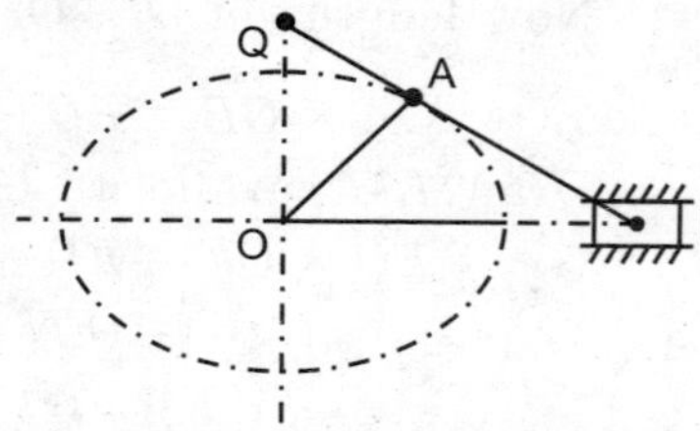

Fig. 4.6. Modified Scott-Russel Mechanism

When point A moves in a circle, the point Q moves along an approximate straight line, if

$$OA = \frac{(AP)^2}{AQ}$$

This is limited to only small displacement of P.

4.7. GRASSHOPPER MECHANISM OF APPROXIMATE STRAIGHT LINE MOTION

This mechanism is a modification of modified Scott-Russel's mechanism. The point P does not slide along a straight line and moves in a circular arc with centre O_1'. The mechanism is shown in Fig. 4.7. It is a four-bar mechanism with all the pairs as turning pairs. The centres O and O_1 are fixed.

The link OA oscillates about O through an angle AOA and causes the pin P to move along a circular are with O_1 as centre and O_1P as radius.

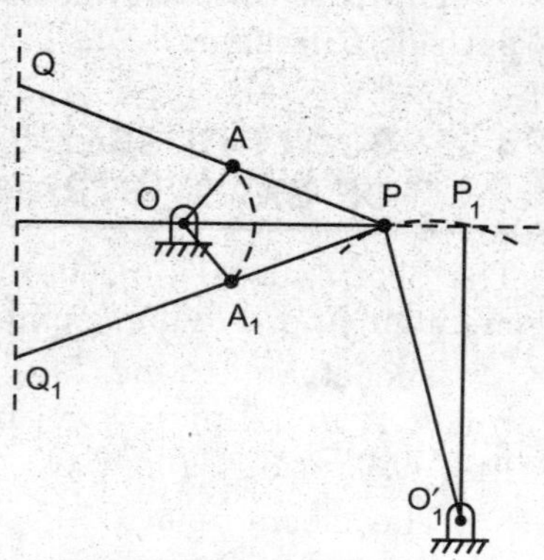

Fig. 4.7. Grasshopper mechanism

For small angular displacement of OP on each side of horizontal, the point Q located on the extension of link PA traces an approximate straight path QQ_1 with the condition that

$$OA = \frac{(AP)^2}{AQ}$$

Note: This mechanism was used as an engine indicator which gave long stroke with a very short crank.

4.8. WATT'S MECHANISM FOR APPROXIMATE STRAIGHT LINE MOTION

The mechanism was developed by Watt for his steam engine to guide the piston rod, in a cylinder to move with approximate straight line motion. It is a crossed four-bar chain mechanism and is shown in Fig. 4.8.

It has four links *OA, AB, BC* and *CO*. The link *CO* is fixed. The link *AO* can oscillate about point *O* and link *BC* can oscillate about point *C*.

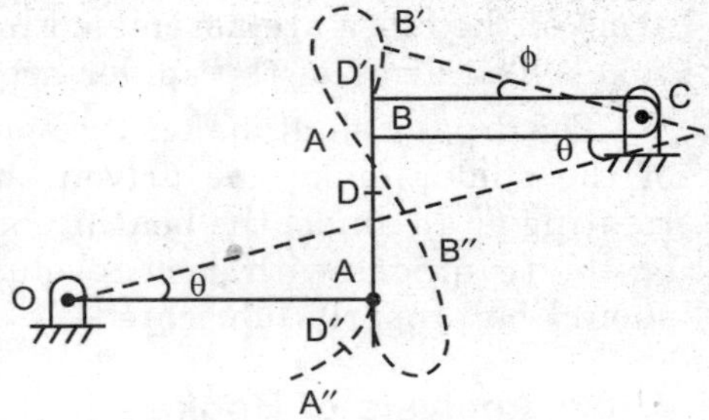

Fig. 4.8. Watt's Mechanism

In the mean position of the mechanism, the links *AO* and *BC* are parallel and link *AB* is perpendicular to *AO* and *BC*. *A* point *D* on the link *AB* will trace an approximate straight line. The three position of point *D* (*D, D′* and *D″*) are on an approximate straight line.

4.9. TCHEBICHEFF MECHANISM

It is a four-bar mechanism and consists of four links *OC, CD, AD* and *OA*. The link *OA* is fixed. The links *OC* and *AD* are equal and crossed. *B* is the midpoint of *CD* and is the tracing point. In extreme position of mechanism, point *D* and *B* lie on the vertical line above *O* and *B* and *C* lie on vertical line above *A*. The mean position is shown by *ADCD* and extreme position by AD_1C_1O.

The locus of point $B(B_1 - B - B_2)$ is a straight line parallel to *OA*.

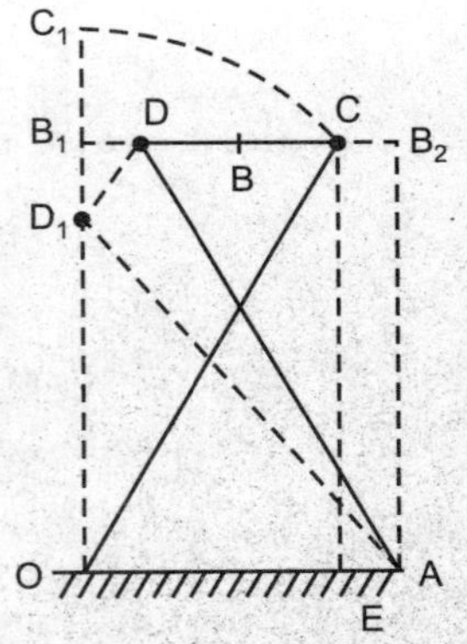

Fig. 4.9. Tchebicheff Mechanism

4.10. HOOKE'S JOINT

A Hooke's joint is a connection between two shafts that permit a driving shaft to drive a driven shaft at an angle. This is also called universal joint. These are used in automobiles because the rear end of the propeller shaft is constantly rising and falling due to:

(*i*) The flexing of the road springs.

(*ii*) They allow for the rear axle assembly to link due to the drive and brake torque reaction.

A Hooke's joint is shown in Fig. 4.10. This joint not only makes a flexible connection between two rigid shafts at an angle but they permit to continue the transmission of power when the angle is continuously changing.

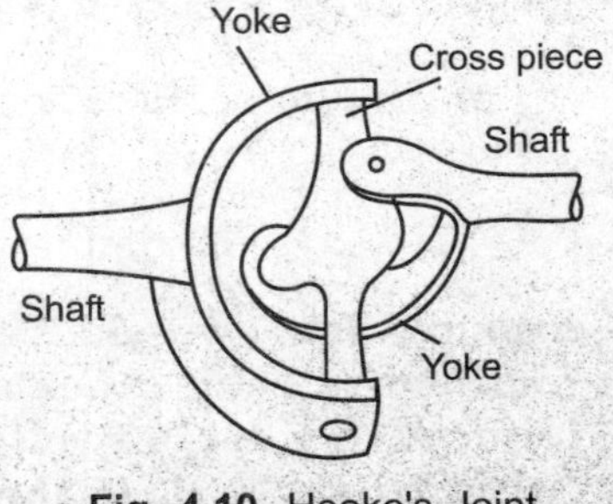

Fig. 4.10. Hooke's Joint

It is a double hinged joint consisting of two Y-shaped yokes. One yoke is provided on the driving shaft and second on the driven shaft. A crossed shape member also known as the spider connects the two yokes. The four arms of the cross are assembled into the bearings in the ends of two shaft yokes. The arms of the spider are also known as trunnions.

The driving shaft makes the spider to notate and the other two trunnions of the spider make the driven shaft to rotate when the two shafts are rotating at an angle, the bearings of the yokes permit them to swing around on the trunnions with each revolution. The bearings of the universal joint should be properly lubricated.

4.10.1 Analysis of Hooke's joint

The driving shaft A and driven shaft B are inclined at an angle α. If the shaft A rotates through an angle θ, Then the velocity ratio

$$\frac{\omega_B}{\omega_A} = \frac{\cos\alpha}{1-\cos^2\theta\sin^2\alpha}$$

Velocity ratio is maximum when $\theta = n\pi$

where $n = 0, 1, 2, 3$.........

Velocity ratio is minimum when $\theta = \frac{n\pi}{2}$

where $n = 1, 2, 3$ ....

$$\left(\frac{\omega_B}{\omega_A}\right)_{max} = \frac{1}{\cos\alpha}$$

$$\left(\frac{\omega_B}{\omega_A}\right)_{min} = \cos\alpha$$

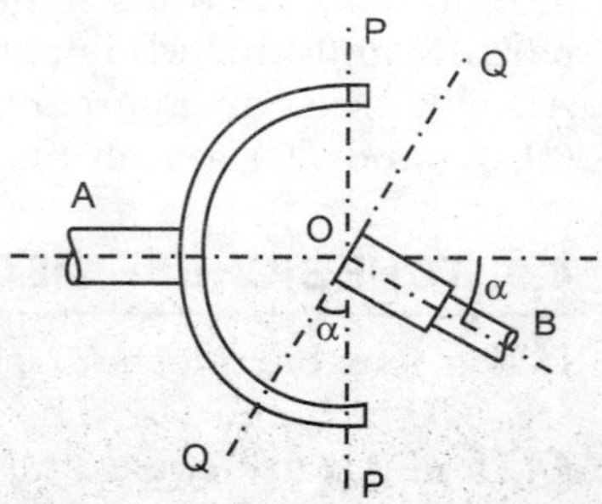

Fig. 4.11. Analysis of Hooke's Joint.

Maximum fluctuations in the velocity ratio, $\frac{\omega_B}{\omega_A}$

$$= \left(\frac{\omega_B}{\omega_A}\right)_{max} - \left(\frac{\omega_B}{\omega_A}\right)_{min} = \frac{\sin\alpha}{\cos\alpha}$$

Angular acceleration of the driven shaft is maximum when

$$\cos 2\theta = \frac{2\sin^2\alpha}{2-\sin^2\alpha}$$

$$\text{Maximum acceleration } \alpha_{max} = \frac{\omega_A^2\cos\alpha\sin^2\alpha\sin 2\theta}{(1-\cos^2\theta\sin^2\alpha)^2}$$

$$\text{Maximum torque} = \frac{W}{g}\,k^2\,(acc)_{max.}$$

Example 4.1. **Two shafts A and B, whose axes are intersecting but inclined to each other at 15° are connected by means of a Hooke's joint. A flywheel of weight 180 kN and radius of gyration 80 mm is fitted to shaft B. If the shaft A rotates at uniform speed of 2000 rpm., what is the maximum torque in B?**

Solution : $\omega_A = \dfrac{2\pi N}{60} = \dfrac{2\pi \times 2000}{60} = 209.44$ rad/s.

For maximum acceleration,

$$\cos 2\theta = \frac{2\sin^2\alpha}{2-\sin^2\alpha} = \frac{2\sin^2 15°}{2-\sin^2 15°}$$

$$\cos 2\theta = 0.0693$$

$$\therefore \quad \theta = 43°$$

$$\text{Maximum acceleration } \alpha_{max} = \frac{\omega_A^2 \cos 15° \sin^2 15° \sin 86°}{(1-\cos^2 43° \sin^2 15°)^2}$$

$$= 3044 \text{ rad/s}^2$$

$$\text{Maximum torque} = \frac{W}{g} k^2 \alpha_{max}$$

$$= \frac{180}{9.8\times 10^3} \times (80)^2 \times 3044$$

$$= 357460 \text{ Nm}$$

$$= \mathbf{357.46 \text{ kN-m Ans.}}$$

4.11. STEERING GEAR MECHANISMS

The steering gear mechanism plays an important role in ensuring easy, comfortable driving all the way from the low speed range to the high speed range.

A correct steering mechanism should ensure pure rolling of all the wheels without slipping under all conditions of vehicle's motion: straight, leftward and rightward. This is possible when the axes of front wheel sub-axle intersect at a point lying on the common axis of rear wheels.

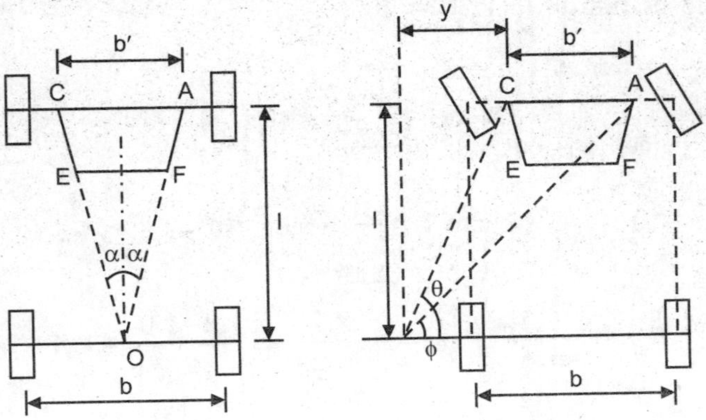

Fig. 4.12. Principle of steering

For correct steering,

$$\cot \phi = \frac{y+b'}{\ell}$$

where ϕ = angle of inside lock

$$\cot\phi = \frac{y}{\ell}+\frac{b'}{\ell} = \cot\theta+\frac{b'}{\ell}$$

where θ = angle of outside lock

$$\therefore \quad \cot\phi - \cot\theta = \frac{b'}{\ell}$$

Turning circle radii for different wheels:

1. For inner front wheel

$$R_{if} = \frac{\ell}{\sin\theta}-\left(\frac{b-b'}{2}\right)$$

2. For outer front wheel,

$$R_{of} = \frac{\ell}{\sin\phi}+\left(\frac{b-b'}{2}\right)$$

3. For the inner rear wheel,

$$R_{ir} = \frac{\ell}{\tan\theta}-\left(\frac{b-b'}{2}\right)$$

4. For the outer rear wheel,

$$R_{or} = \frac{\ell}{\tan\phi}+\left(\frac{b-b'}{2}\right).$$

Example 4.2. **The wheel base of a car is 2.7 m and pivot centres are 1 m. The wheel track is 1.2 m. Calculate the correct angle of outside lock and turning circle radii of the outer front wheel and inner rear wheel when angle of inside lock is 40°.**

Solution:

Wheel base, ℓ = 2.7 m

Pivot centre, b' = 1.0 m

Wheel track, b = 1.2 m

Angle of inside lock, θ = 40°

(*i*)

$$\cot\phi - \cot\theta = \frac{b'}{\ell}$$

$$\cot\phi - \cot 40° = \frac{1.0}{2.7}$$

$$\cot\phi = \frac{1.0}{2.7} + \cot 40°$$

$$\phi = 32.6°$$

(*ii*)

$$R_{of} = \frac{\ell}{\sin\phi}+\frac{b-b'}{2} = \frac{2.7}{\sin 32.6°}+\frac{1.2-1.0}{2} = 5.11 \text{ m}$$

(*iii*)

$$R_{ir} = \frac{\ell}{\tan\theta}-\frac{b-b'}{2} = \frac{2.7}{\tan 40°}-\frac{1.2-1.0}{2} = \mathbf{3.12\ m}$$

Example 4.3. **The ratio between the width of the front axle and that of wheel base of a steering mechanism is 0.44. At the instant when the front inner wheel is turned by 18°, what should be the angle turned by outer front wheel for perfect steering? *[UPTU 2010-11]***

Solution :

$$\frac{b'}{\ell} = 0.44$$

$$\theta = 18°$$

$$\cot \phi - \cot \theta = \frac{b'}{\ell}$$

$$\therefore \quad \cot \phi - \cot 18° = 0.44$$

$$\cot \phi = 0.44 + 3.078 = 3.518$$

$$\therefore \quad \phi = 15.9° \textbf{ Ans.}$$

4.12. TYPES OF STEERING MECHANISM

Main types of steering mechanisms are:

1. Davis Steering gear

This mechanism satisfies the requirements of accurate steering under all conditions of motion. However, it contains sliding pairs resulting in friction and wearing. Excessive wearing can cause inaccuracy in steering after some time.

2. Ackermann Steering gear

This mechanism has only turning pairs and therefore it is widely used. However, this steering gear does not satisfy the fundamental equation of accurate steering at all positions. It fulfills the fundamental equation of accurate steering only at the middle and extreme positions.

4.12.1. Davis Steering Gear

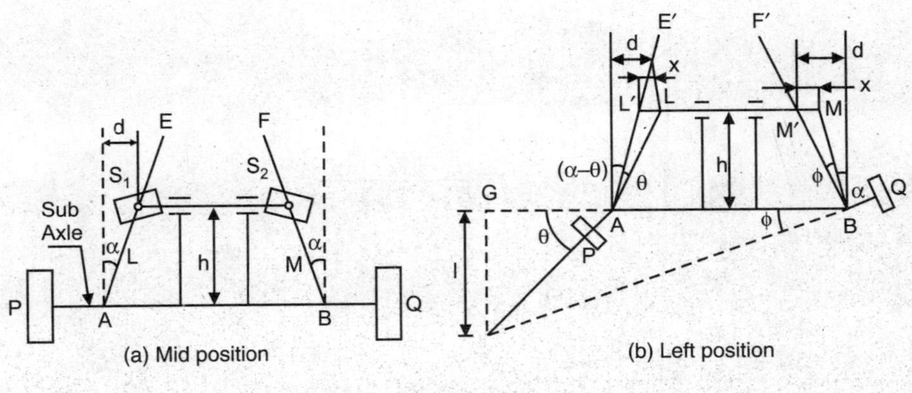

Fig. 4.13. Davis steering gear

It has two arms AE and BF fixed to stub axle AP and BP. It transforms into two similar bell-crank levers PAE and QBF pivoted at A and B respectively. Cross link LM is constrained to slide parallel to AB, sliders S_1 and S_2 are pinned to the ends of cross link LM and are free to slide on links AE and BF.

The cross-link LM is moved to right or left to affect steering

Assure d = Horizontal distance between A and L
= Horizontal distance between B and M.
h = Vertical distance between AB and LM
x = Distance moved by cross link from mid-position
= $MM' = LL'$
α = Angle of inclination of AE or BF with the vertical when vehicle is in mid-position
θ, ϕ = angles turned by stub axles.

Now $$\tan(\alpha - \theta) = \frac{d-x}{h}$$

or $$\frac{\tan\alpha - \tan\theta}{1+\tan\alpha\tan\theta} = \frac{d-x}{h}$$

$$\tan\alpha = \frac{d}{h}$$

$\therefore$ $$\frac{\frac{d}{h} - \tan\theta}{1+\frac{d}{h}\tan\theta} = \frac{d-x}{h}$$

or $$\frac{d-h\tan\theta}{1+d\tan\theta} = \frac{d-x}{h}$$

$$h(d - h\tan\theta) = (d-x)(h + d\tan\theta)$$

$$hd - h^2\tan\theta = hd + d^2\tan\theta - xh - xd\tan\theta$$

$$xh = (d^2 - xd + h^2)\tan\theta$$

$\therefore$ $$\tan\theta = \frac{xh}{(d^2 - xd + h^2)}$$

Now $$\tan(\alpha + \phi) = \frac{d+x}{h}$$

$\therefore$ $$\tan\phi = \frac{hx}{(d^2 + xd + h^2)}$$

For correct steering

$$\cot\phi - \cot\theta = \frac{b'}{\ell}$$

$$\frac{d^2+xd+h^2}{hx} - \frac{d^2-xd+h^2}{hx} = \frac{b'}{\ell}$$

$$\frac{2xd}{hx} = \frac{b'}{\ell}$$

$$\frac{d}{h} = \frac{b'}{2\ell}$$

$\therefore$ $$\tan\alpha = \frac{b'}{2\ell}$$

Example 4.4. **A vehicle is moving in a straight line. The distance between the pivots of the front axle of a Davis steering gear is 1.25 m. The wheel base is 2.7 m. Find the inclination of the track arm to the longitudinal axis of the vehicle.**

Solution : Distance between points, $b' = 1.25$ m

The wheel base, $\ell = 2.7$ m

Let α = inclination of the track arm to the longitudinal axis

$$\tan \alpha = \frac{b'}{2\ell} = \frac{1.25}{2 \times 2.7} = 0.2315$$

$$\alpha = \tan 0.2315 = \mathbf{13.03° \text{ Ans.}}$$

4.12.2. Ackermann Steering Gear

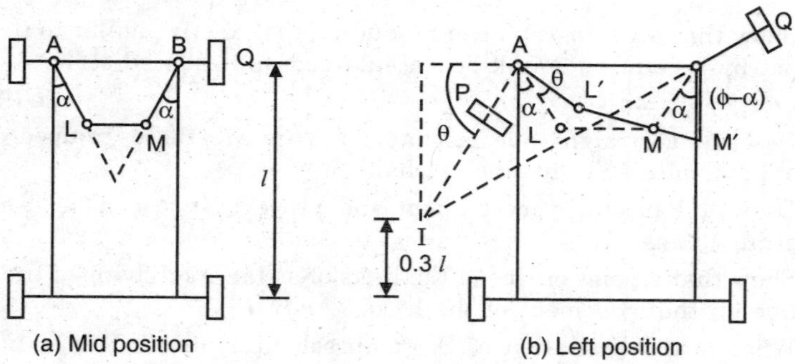

Fig. 4.14. Ackermann Steering Gear

The mechanism consists of four turning gears *AL, LM, MB* and *BA*. The arms *AL* and *BM* are of equal length and are connected by hinge joints with the axle at *A* and *B* respectively. The links *AB* and *LM* are of unequal lengths.

The steering gear satisfies the fundamental equation of correct steering in the following positions.

1. Vehicle moving along straight path
2. Vehicle is steering to the left at correct angle
3. Vehicle is steering to the right at correct angle

In all other position, the fundamental equation of correct steering does not hold good.

The instantaneous centre *I* lies on a line parallel to rear axis approximately 0.3ℓ from it whereas in Davis steering gear it lies on the rear axis. The whole of mechanism of the Ackermann gear is on the back of front wheels whereas in Davis gear, it is in front of the wheels.

Assume that the projects of *LM* and *L′M′* on *AB* are equal

$$AL' \cos[90° - (\alpha + \theta)] - AL \cos(90° - \alpha)] = BM \cos(90° - \alpha) + BM' \sin(\phi - \alpha)$$

But $AL' = AL = BM = BM'$

$$\therefore \quad \sin(\alpha + \theta) - \sin \alpha = \sin \alpha + \sin(\phi - \alpha)$$

or $\sin \alpha \cos \theta + \cos \alpha \sin \theta - \sin \alpha = \sin \alpha + (\sin \phi \cos \alpha - \cos \phi \sin \alpha)$

or $\sin\alpha\cos\theta - \sin\alpha - \sin\alpha + \sin\alpha\cos\phi = \cos\alpha\sin\phi - \cos\alpha\sin\theta$

$\sin\alpha(\cos\theta + \cos\phi - 2) = \cos\alpha(\sin\phi - \sin\theta)$

$$\frac{\sin\alpha}{\cos\alpha} = \frac{\sin\phi - \sin\theta}{\cos\theta + \cos\phi - 2}$$

$$\tan\alpha = \frac{\sin\phi - \sin\theta}{\cos\theta + \cos\phi - 2}$$

For correct steering values of θ and ϕ are known, therefore value of α can be calculated.

QUESTION BANK

1. Show that a pantograph can produce paths exactly similar to the ones traced out by a point on a link on an enlarged and reduced scale.
   [UPTU 2010-11]
2. Describe a straight line mechanism. Why are these grouped as exact and approximate straight line mechanism?
3. Draw a Pauceillier mechanism and prove that it can be used to trace a straight line.
4. Show that a point on one of the links of a Hart's mechanism traces a straight line on the movement of the links.
5. What is the limitation of Scott-Russel mechanism? How is it modified to overcome the limitation?
6. How is Grass-Hopper mechanism derived from modified Scott-Russel mechanism?
7. Write notes on the following:
   (*a*) Watt mechanism.
   (*b*) Tchebicheff mechanism
8. Derive an expression for velocity ratio of a Hooke's joint.
9. What is fundamental equation of steering gear? What steering gear fulfils the conditions?
10. Show that an Ackermann steering gear does not satisfy the fundamental conditions of steering gear at all positions. Why is it widely used?

NUMERICAL PROBLEMS

1. The driving shaft of a Hooker's joint has uniform angular velocity of 280 rpm. Determine the maximum permissible angle between the axes of the shafts to permit a maximum variation in velocity of driven shaft by 8% of mean velocity.
   [Ans: 22.6°]
2. In a Davis steering gear, the distance between steering pivots is 1.35 m and that between car axle is 2.4 m. Determine the inclination of the track arms to the longitudinal axis of the car when the car moves along a straight line.
   [Ans: 15°42′]

❑❑❑

UNIT–III

5. BEARING FRICTION

6. BELT AND PULLEY DRIVE

7. BRAKES AND DYNAMOMETERS

Chapter 5

BEARING FRICTION

5.1. INTRODUCTION

Whenever the surfaces of two solid bodies are in contact, there is some resistance to sliding between them. The opposing force to the movement is called friction or force of friction. It is due to interlocking of surfaces as a result of the presence of some roughness and irregularities at the contact surfaces. The resisting force acts in a direction opposite to the movement. A force of friction comes into play whenever there is relative motion between two parts. Some energy is wasted in order to overcome the resistance due to force of friction.

Friction is both desirable and undesirable in engineering applications. There are many devices in engineering which work due to friction. Lifting machines such as screw jack, power transmission by belts and ropes, friction clutches, brakes, nuts and bolts, jib and cotter joints, etc, work due to friction. The friction should be utilized by the use of proper materials and shape of components. The vehicles can run on roads and inclined surfaces due to friction only. A man cannot walk if there is no friction.

However, there is resistance and wastage of energy in the bearings. Friction causes loss of energy in moving parts of machines and heating of bodies and may affect life of the machines. The efficiency of the machines decreases due to presence of friction.

Solid, liquid and gaseous lubricants are used to reduce friction by separating the moving surfaces.

A body at rest tending to move encounters static friction whereas force of friction encountered by a body in motion is called dynamic or kinetic friction. The friction can be sliding friction on flat surfaces or rolling friction when one body rolls over other or surfaces are separated by balls and rollers. The rolling friction is always less than sliding friction when surfaces slide one over the other.

The friction that exists when no lubricant is used is called dry friction or solid friction. Laws of static and dynamic friction are based on experiments conducted by Prof. Coulomb and are called laws of Coulomb friction.

5.2. DRY FRICTION

Force of friction or frictional force is the opposing force to the movement of one body over the surface of another body.

A block of weight W is resting on a rough horizontal surface and is being pulled by a horizontal force P.

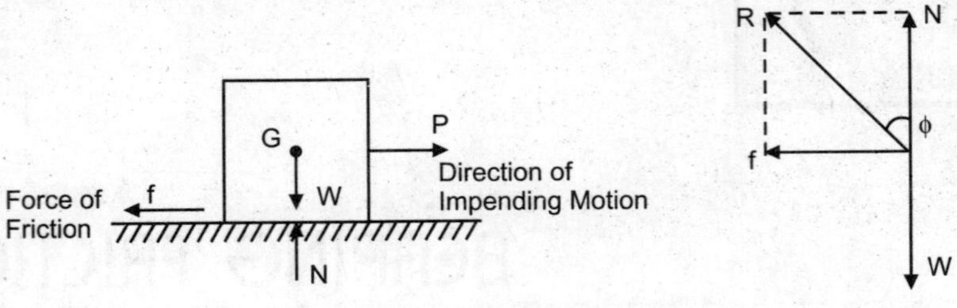

Fig. 5.1. Force of dry friction

Assures, W = Weight of the body [N]

$= mg$

N = Normal reaction [N]

f = Force of friction [N]

P = Force applied to the body [N]

R = Total reaction [N]

ϕ = Angle of friction

The force of friction (f) always acts in a direction opposite to the motion and opposes the tractive effort (P). It is a reactive or passive force and exists as long as tractive effort is present. The friction force adjusts itself depending upon the magnitude of tractive effort or applied force.

1. Coefficient of Friction

The force of friction is directly proportional to the normal reaction between the surface.

$$f \propto N$$

or
$$\frac{f}{N} = \text{constant} = \mu$$

μ is called coefficient of friction and depends upon the nature of contact surfaces.

$$f = \mu N.$$

2. Angle of Friction

R is the resultant of normal force N and force of friction f.

$$R = \sqrt{N^2 + f^2}$$

Angle of friction (ϕ) is the angle made by resultant (R) with the normal reaction (N)

$$\tan \phi = \frac{f}{N}$$

or
$$\phi = \tan^{-1} \frac{f}{N}$$

$$= \tan^{-1} \mu$$

$\therefore$
$$\mu = \tan \phi$$

3. Limiting friction

The friction force, f increases to oppose P until a maximum value of f_m is reached. If P is further increased, the body starts sliding as the force of friction cannot balance the force P. The value of f is lowered from f_m, which is the limiting or maximum friction force that is exerted at the time when motion is about to begin.

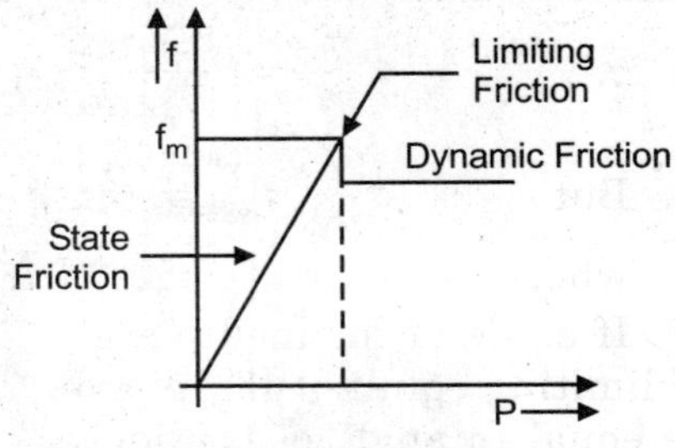

Fig. 5.2. Limiting friction

4. Core of friction

Let $OA = N$ = Normal reaction

$OC = f_m$ = Limiting friction force

$OB = R$ = Resultant force

$\angle AOB = \phi$ = Max. anlge of friction

If vector OB is revolved about vertical axis OA, the generated cone OBD is called cone of friction.

If the force of friction $f(AE)$ and resultant OE fall within the cone of friction, there will be no motion of the body. This priciple is used in self-locking mechanisms of lifting machines, i.e., screw jack and taper pins.

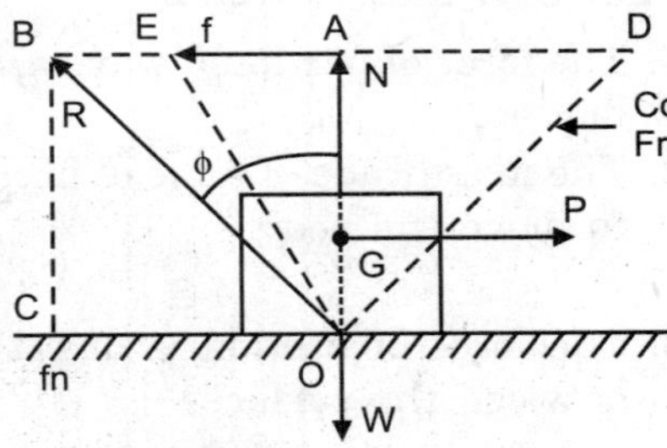

Fig. 5.3. Core of friction

5. Angle of repose

A body of weight W is lying on a rough plane inclined at an angle α with the horizontal. The body is in equilibrium under the action of following forces.

(*i*) Weight of the body W acting vertically downwards. It has two components, *i.e.*, $W \sin \alpha$ parallel to inclined plane and $W \cos \alpha$ normal to plane.

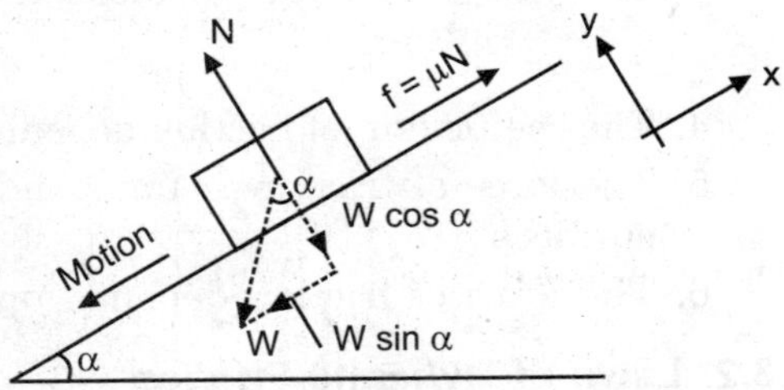

Fig. 5.4. Angle of repose

(*ii*) Normal reaction, N acting in a direction normal to inclined plane.

(*iii*) Friction force, f acting in a direction apposite to motion. When the body tends to slide down the plane, the friction force must act up the plane and when the body is being pulled up, the force of friction acts downward to oppose the motion.

Selecting the reference coordinate system with x-axis in the direction of inclined plane and y-axis normal to inclined plane and applying condition of equilibrium,

$$\Sigma F_x = 0 \qquad f = W \sin \alpha$$

$$\Sigma F_y = 0 \qquad N = W \cos \alpha$$

$$\therefore \quad \frac{f}{N} = \frac{W \sin\alpha}{W \sin\alpha} = \tan\alpha$$

But $$\frac{f}{N} = \mu = \tan\phi$$

where ϕ = angle of friction

If angle of inclined place, α is altered so that the body is in the state of limiting equilibrium, *i.e.*, on the verge of sliding down, the angle α will be equal to angle of friction ϕ. This value of α is called angle of repose.

The angle of repose may be defined as the angle of inclined plane on which the body is in the state of limiting equilibrium. The angle of repose is equal to angle of friction.

5.3. COULOMB LAWS OF FRICTION

5.3.1. Laws of Static Friction

1. The force of friction always acts opposite to the direction of impending motion.
2. The magnitude of force of friction is equal to the applied force tending to move the body.
$$f = P$$
3. The force of friction is directly prepositional to the normal reaction between the surfaces
$$f \propto N$$
or $$\frac{f}{N} = \text{constant} = \mu$$
$$\therefore \quad f = \mu N$$
4. The coefficient of friction depends upon the nature of contact surfaces.
5. The force of friction is independent of the area of contact of the two surfaces.
6. The force of friction depends upon the roughness of the surfaces.

5.3.2. Laws of Dynamic Friction

1. The force of friction always opposes the motion of the body and acts in the direction opposite of motion.
2. The force of friction is directly proportional to the normal reaction between the contacting surfaces in relative motion.
3. The force of friction is slightly less than that for state friction, *i.e.*, the coefficient of friction is less for dynamic friction than for limiting friction.
4. For low velocities, total amount of friction force that can be developed is practically independent of velocity.

5.4. FRICTION ON ROUGH INCLINED PLANE

Inclined plane is very frequently used in the design of various components of machines and mechanism such as screw jack, wedge, screw threads used for nuts and bolts, etc. It is therefore, important to study the relationship between various forces acting on a body when it slides either up or down an inclined place. It is also necessary to study the relationship of forces acting on vehicles moving along the inclined plane.

5.4.1. Body at Rest

When the angle of inclination of inclined plane is less than angle of friction, the body will be in equailibrium without motion. Draw free body diagram and apply equilibrium conditions.

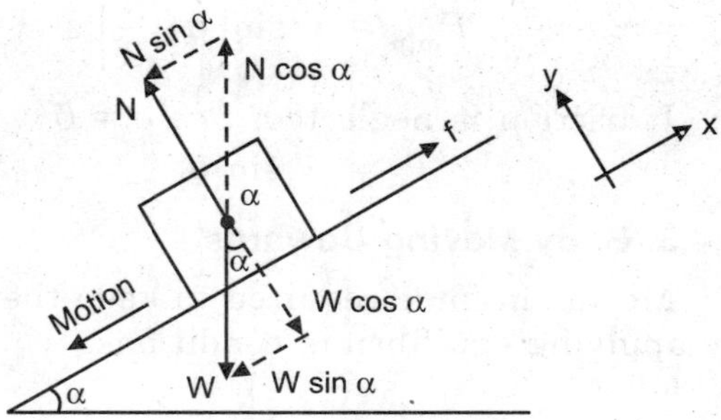

Fig. 5.5. Body at Rest

$\Sigma F_x = 0 \qquad f = \mu N = \text{N} \sin \alpha$

$\Sigma F_y = 0 \qquad W = N \cos \alpha$

or $\qquad f = \mu N = W \sin \alpha$

$$N = W \cos \alpha$$

$$\therefore \qquad \tan \alpha = \frac{f}{N} = \mu = \tan \phi$$

$$\alpha = \phi$$

The maximum angle of inclination of plane with the horizontal when body starts sliding of its own is equal to angle of repose or limiting angle of friction.

5.4.2. Body Sliding Downwards

When a body of weight W is on the verge of sliding downwards, a minimum force P is required to restrict the motion.

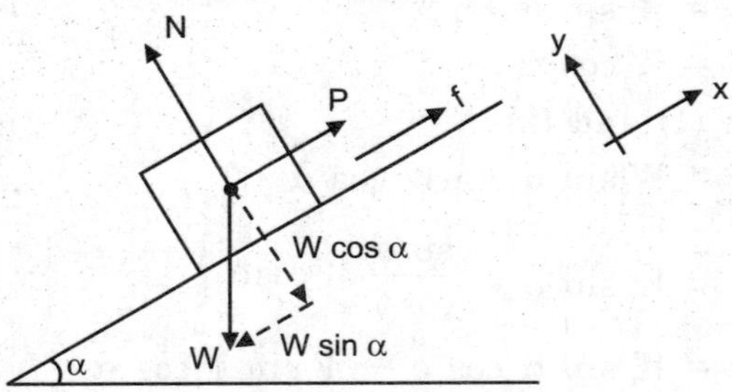

Fig. 5.6. Body Sliding downwards

Free body diagram is shown in Fig. 5.6. Applying condition of equilibrium

$\Sigma F_x = 0 \qquad f = \mu N = W \sin \alpha - P$

or $\qquad P = W \sin \alpha - \mu N \qquad$ ...(1)

$\Sigma F_y = 0 \qquad N = W \cos \alpha \qquad$ ...(2)

From equations (1) and (2)

$$P = W \sin \alpha - \mu W \cos \alpha$$

$$= W\left(\sin \alpha - \frac{\sin \phi}{\cos \phi} \cos \alpha \right) \qquad \left[\because \ \mu = \tan \phi = \frac{\sin \phi}{\cos \phi} \right]$$

$$\therefore \qquad P \cos \phi = W(\sin \alpha \cos \phi - \cos \alpha \cos \phi)$$

$$P = \frac{W\sin(\alpha - \phi)}{\cos\phi}$$

For a given value of α, P is minimum when denominator is maximum, *i.e.*, $\cos \phi = 1$ or $\phi = 90°$.

$$P_{min} = W\sin\left(\alpha - \frac{\pi}{2}\right)$$

If friction is neglected, *i.e.*, $\phi = 0$

$$P_0 = W \sin \alpha$$

5.4.3. Body Moving Upwards

Minimum force required to keep the body in equilibrium can be found by applying equilibrium conditions.

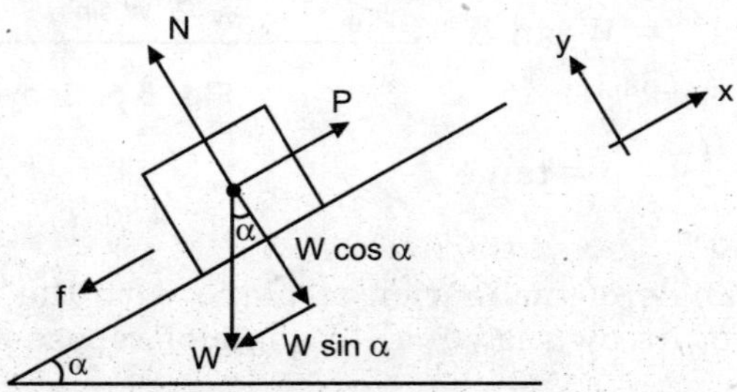

Fig. 5.7. Body moving upwards

$\Sigma F_x = 0$ $\quad f = P - W \sin \alpha$

$$P = W \sin \alpha + f$$

$$= W \sin \alpha + \mu N \quad ...(1)$$

$\Sigma F_y = 0$ $\quad N = W \cos \alpha \quad ...(2)$

From equations (1) and (2)

$$P = W \sin \alpha + \mu W \cos \alpha$$

$$= W \sin \alpha + \frac{\sin\phi}{\cos\phi} \cdot W\cos\alpha \qquad \left[\because \mu = \tan\phi = \frac{\sin\phi}{\cos\phi}\right]$$

$$P \cos \alpha = W \sin \alpha \cos \phi + W \sin \phi \cos \alpha$$

$$= W \sin (\alpha + \phi)$$

$$\therefore \qquad P = \frac{W\sin(\alpha + \phi)}{\cos\phi}$$

For a given value of α, P is minimum when denominator is maximum, *i.e.*, $\cos \phi = 1$ or $\phi = 90°$.

$$P_{min} = W\sin\left(\alpha + \frac{\pi}{2}\right)$$

If friction is neglected,

i.e., $\phi = 0$

$$P_o = W \sin \alpha.$$

5.5. EFFICIENCY OF INCLINED PLANE

The efficiency of an inclined plane is the ratio of force required neglecting friction to the force required taking into account the friction to move the body on an inclined plane.

$$\eta = \frac{P_o}{P}$$

1. Body sliding downwards

The value of P will be less than P_o.

$$\therefore \qquad \eta = \frac{P_o}{P} = \frac{W\sin(\alpha-\phi)}{\cos\phi}\times\frac{1}{W\sin\alpha}$$

$$\text{or} \qquad \eta = \frac{\sin(\alpha-\phi)}{\cos\phi\sin\alpha}$$

2. Body moving upwards

$$\eta = \frac{P_o}{P} = \frac{W\sin\alpha\times\cos\phi}{W\sin(\alpha-\phi)}$$

$$= \frac{\sin\alpha\cos\phi}{\sin(\alpha-\phi)}.$$

Example 5.1. **A body of weight W is placed on a rough inclined plane having inclination α with the horizontal. The force P is applied horizontally to drag the body. If the body is on the point of motion up the plane, prove that P is given by.**

$$\mathbf{P = W\tan(\alpha+\phi)}$$

where $\quad \phi$ **= angle of friction.**

Solution : Draw the free body diagram and apply equilibrium conditions

$\Sigma F_x = 0 \qquad P\cos\alpha = W\sin\alpha + f = W\cos\alpha + \mu N$ ...(1)

$\Sigma F_y = 0 \qquad N = W\cos\alpha + P\sin\alpha$ ...(2)

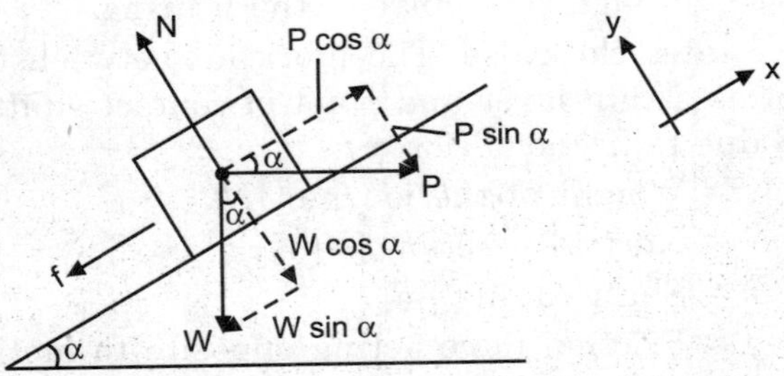

From equations (1) and (2)

$$P\cos\alpha = W\sin\alpha + \mu[W\cos\alpha + P\sin\alpha]$$

$$= W\sin\alpha + \mu W\cos\alpha + \mu P\sin\alpha$$

$$P\cos\alpha - \mu\,P\sin\alpha = W\cos\alpha + \mu\,W\cos\alpha$$

$$\therefore \qquad P = \frac{\sin\alpha+\mu\cos\alpha}{\cos\alpha-\mu\cos\alpha}$$

But $\qquad \mu = \tan\phi \qquad$ (By definition)

$$\therefore \qquad P = W\left[\frac{\sin\alpha+\tan\phi\cos\alpha}{\cos\alpha-\tan\phi\sin\alpha}\right]$$

$$P = W\left[\frac{\sin\alpha\cos\phi + \sin\phi\cos\alpha}{\cos\alpha\cos\phi - \sin\alpha\sin\phi}\right]$$

$$= W\frac{\sin(\alpha+\phi)}{\cos(\alpha+\phi)}$$

$$\therefore \qquad P = W\tan(\alpha + \phi)$$

5.6. FRICTION IN JOURNAL BEARING–FRICTION CIRCLE

A journal bearing consists of a rotating shift called journal and a fixed casing called a bearing. The radius of the journal is kept less than that of the bearing as per running fit tolerances.

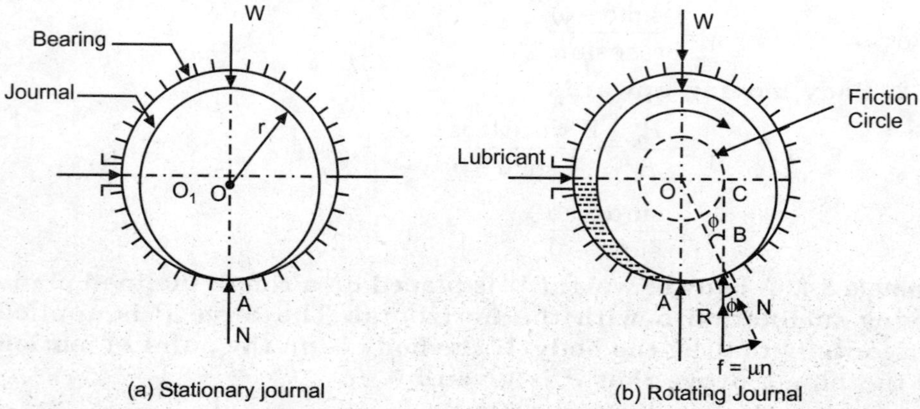

Fig. 5.8. Journal Bearing

(*a*) The shaft is at rest inside the bearing and there is no lubricant. The weight *W* of the journal acts at point *A* and bearing offers a normal reaction *N* upwards. Point *A* is called seat or point of pressure. Both *W* and *N* act through the centre of the bearing.

(*b*) The shaft rotates clockwise. The lubricant between the journal and bearing forms a thin layer and point of contact shifts rightwards to position *B* due to frictional force *f*.

Assures, W = Weight of the journal [*N*]

N = Normal reaction [*N*]

r = Shaft radius [m]

f = Friction force acting opposite to motion [N]

$= \mu N$

ϕ = Angle between R and N

R = Resultant of N and f

T = Frictional torque [*Nm*]

$= fr$

In order to maintain the rotation, there must be a couple rotating the shaft. For uniform motion, the dynamic resulting force acting on the shaft must be zero and the resultant turning moment must be zero.

$$\therefore \quad R = W$$

$$T = W \times OC$$

$$= W \times OB \sin \phi$$

$$= Wr \sin \phi$$

For small ϕ, $\sin \phi = \tan \phi = \mu$

$$\therefore \quad T = Wr \tan \phi = Wr\mu \ [N]$$

The power consumed in friction,

$$P = \frac{2\pi NT}{60 \times 1000} \ [\text{kW}]$$

where N = speed of the shaft in rpm

Note : If a circle is drawn with centre O and radius OC (= $r \sin \phi$), this circle is called the friction circle of bearing. The resultant force R acts along a tangent to the friction circle.

***Example* 5.2. A shaft of diameter 60 mm is running in a journal bearing at 1440 rpm. The bearing carries a load of 2000 *N* and coefficient of friction of lubricant is 0.03. Find the power transmitted.**

Solution :

$$d = 60 \text{ mm}$$

$$r = 30 \text{ mm} = 0.03 \text{ m}$$

$$W = 2000 \text{ N.}$$

$$\mu = 0.03$$

$$N = 1440 \text{ rpm}$$

The torque transmitted

$$T = \text{W}.r.\mu$$

$$= 2000 \times 0.03 \times 0.03$$

$$= 1.8 \text{ Nm}$$

The power transmitted,

$$P = \frac{2\pi NT}{60 \times 1000}$$

$$= \frac{2\pi \times 1.8 \times 1440}{60 \times 1000}$$

$$= \mathbf{0.271 \text{ kW Ans.}}$$

5.7. FRICTION IN PIVOT AND COLLAR BEARINGS

The propeller shafts of ships, vertical shafts of hydraulic turbines, horizontal shafts of steam and gas turbines carry axial thrust which has to be supported by pivot and collar bearings. These are placed at the end of the shaft and can have different shapes.

The study of friction in pivot and collar bearings is based on the following assumptions:

1. Uniform pressure distribution on the bearing surface.
2. Uniform wear of the bearing surface.

The most widely used shapes of pivot and collar bearings are shown below.

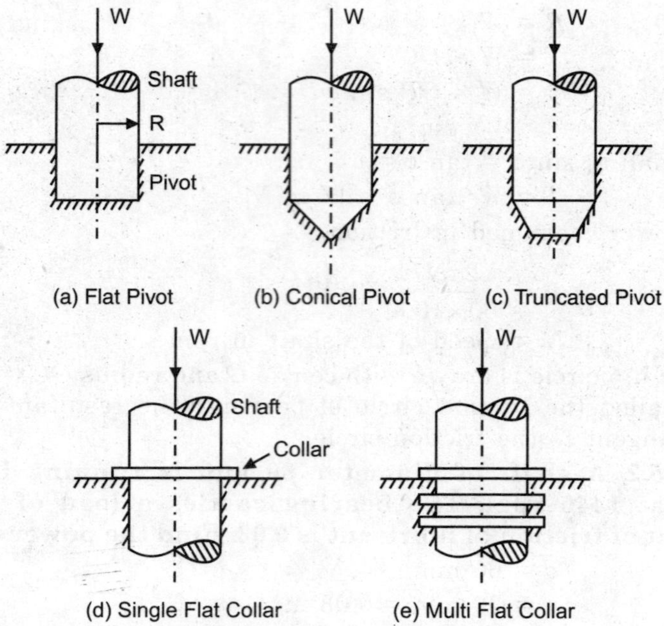

Fig. 5.9. Pivot and Collar Bearings

5.7.1. Flat Pivot Bearing

A flat pivot bearing is also called footstep bearing and supports the axial load of a vertical shaft. There will be sliding friction along the surface of contact between the shaft and the bearing.

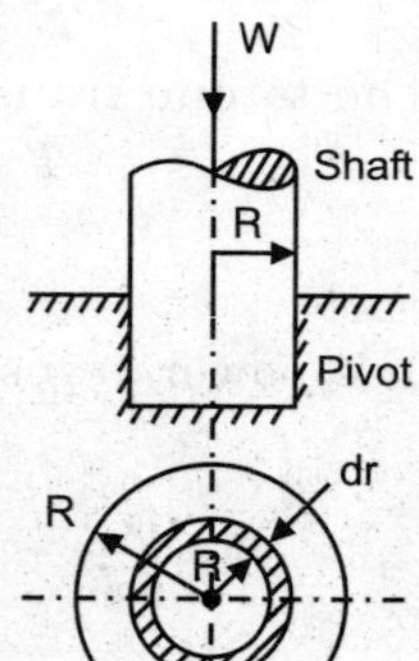

Fig. 5.10. Flat Pivot Bearing

Assure, W = Load transmitted over the bearing surface

R = Radius of bearing surface

p = Pressure on the bearing surface

$$= \frac{W}{\pi R^2}$$

μ = Coefficient of friction

(*a*) Case of Uniform Pressure

Consider an elementary ring of radius r and thickness dr.

Load on the ring = Area × Pressure

$$\delta W = p \times 2\pi r\ dr.$$

$$W = \int_O^R 2\pi rp\ dr = 2\pi p \left|\frac{r^2}{2}\right|_0^R = \pi p R^2$$

$$\therefore \qquad p = \frac{W}{\pi R^2}.$$

Frictional resistance on the elementary ring acting tangentially at radius r,

$$f = \mu\ \delta W = \mu p \times 2\pi r\ dr = 2\pi\mu pr\ dr.$$

Frictional torque on the ring.

$$\delta T = fr = 2\pi\mu pr^2 dr$$

Total frictional torque on the bearing

$$T = \int_0^R 2\pi\mu pr^2 dr = 2\pi\mu p \int_0^R r^2 dr$$

$$= 2\pi\mu p \left|\frac{r^3}{3}\right|_0^R = \frac{2}{3}\pi\mu\ pR^3$$

$$= \frac{2}{3}\mu WR \qquad \left[\because p = \frac{W}{\pi R^2}\right]$$

The power lost in friction,

$$P = \frac{2\pi NT}{60 \times 1000}\ [\text{kW}]$$

where N = speed of the shaft in rpm

(*b*) Case of uniform wear

For uniform wear of bearing,

$$pr = \text{constant} = C$$

$$\therefore \qquad p = \frac{C}{r}.$$

Load on the elementary ring,

$$\delta W = p \times 2\pi r dr$$

$$= \frac{C}{r} \times 2\pi r dr = 2\pi C dr$$

Total load on the bearing

$$W = \int_O^R 2\pi C\ dr = 2\pi CR$$

$$\therefore \qquad C = \frac{W}{2\pi R}$$

Frictional torque on the ring,

$$\delta T = 2\pi\mu pr^2\ dr = 2\pi\mu \frac{C}{r} \times r^2 dr$$

$$= 2\pi\mu Cr\ dr.$$

Total frictional torque on the bearing,

$$T = \int_0^R 2\pi\mu Cr\ dr = 2\pi\mu C \left|\frac{r^2}{2}\right|_0^R$$

$$= 2\pi\mu CR^2$$

$$C = \frac{W}{2\pi R}$$

$$\therefore \quad T = \frac{1}{2}\mu WR$$

Example 5.3. **A vertical shaft of 150 mm carriers a load of 20 kN and runs at 100 rpm. The shaft is supported by a flat footstep bearing with a coefficient of friction 0.05. Estimate the power lost in friction assuming uniform pressure distribution.**

Solution :

$$D = 150 \text{ mm}$$
$$R = 75 \text{ mm} = 0.075 \text{ m}$$
$$N = 100 \text{ rpm}$$
$$W = 20 \text{ kN} = 20 \times 10^2 \text{ N}$$
$$\mu = 0.05$$

The total frictional torque,

$$T = \frac{2}{3}\mu WR = \frac{2}{3} \times 0.05 \times 20 \times 10^3 \times 0.075$$
$$= 50 \text{ Nm}$$

Power lost in friction,

$$P = \frac{2\pi NT}{60 \times 1000} = \frac{2\pi \times 100 \times 50}{60 \times 1000} = \mathbf{0.5235\ kW\ Ans.}$$

5.7.2. Conical Pivot Bearing

A conical pivot bearing suppoting a shaft of load W is shown in Fig. 5.11.

Assure, R = Radius of shaft

α = Semi-angle of cone

μ = Cofficient of friction between shaft and bearing

p_n = Normal Ressure

Consider a small ring of radius r and thickness dr. It has a length dl along the cone.

$$dl = dr \text{ cosec } \alpha$$

Area of ring,

$$A = 2\pi r dl = 2\pi r dr \text{ cosec } \alpha$$

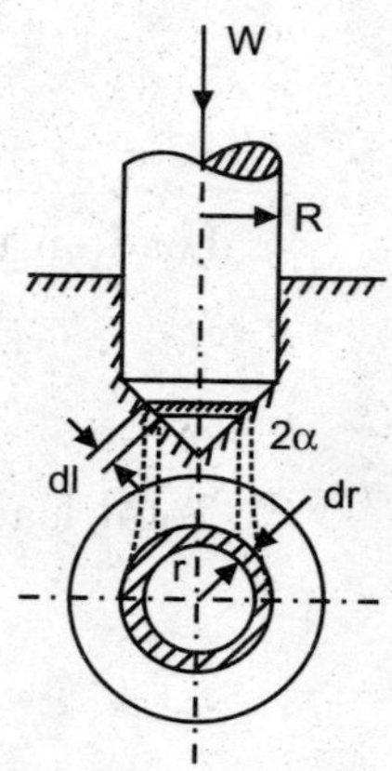

Fig. 5.11. Conical Pivot Bearing

(*a*) Case of uniform pressure

Normal load acting on the bearing,

$$\delta W_n = p_n \times A = p_n \times 2\pi r dr \text{ cosec } \alpha$$

Vertical load acting on the bearing,

$$\delta W = \delta W_n \sin \alpha = p_n \times 2\pi r dr \text{ cosec } \alpha \times \sin \alpha$$
$$= p_n \times 2\pi r\, dr,$$

Total vertical load on the bearing,

$$W = \int_0^R p_n \times 2\pi r dr = 2\pi p_n \left|\frac{r^2}{2}\right|_0^R$$

$$W = \pi R^2 p_n$$

$$\therefore \quad p_n = \frac{W}{\pi R^2}$$

Force of friction on the ring acting tangentially at radius r,

$$f = \mu \delta W_n = \mu p_n \times 2\pi r dr \text{ cosec } \alpha$$
$$= (2\pi \mu p_n \text{ cosec } \alpha)\, r dr$$

Frictional torque on the ring,

$$\delta T = f \times r = (2\pi\mu\, p_n \text{ cosec } \alpha)\, r^2 dr_n$$

$\therefore$ Total frictional torque on the bearing,

$$T = \int_0^R \delta T = \int_0^R (2\pi\mu p_r \alpha \operatorname{cosec} \alpha) r^2 dr_n$$

$$= 2\pi\mu\, p_n \text{ cosec } \alpha \left|\frac{r^3}{3}\right|_0^R$$

$$= \frac{2}{3}\pi\mu p_n \operatorname{cosec} \alpha\, R^3$$

Now $$p_n = \frac{W}{\pi R^2}$$

$$\therefore \quad T = \frac{2}{3}\mu W.R \operatorname{cosec} \alpha$$

$$= \frac{2}{3}\mu W \ell \qquad [\because \ell = R \text{ cosec } \alpha]$$

(b) Case of uniform wear

For uniform wear of bearing

$$pr = C$$

or $$p = \frac{C}{r}$$

Vertical load on the ring

$$\delta W = 2\pi r dr p = \frac{C}{r} \times 2\pi r dr = 2\pi C dr$$

Total load on the bearing

$$W = \int_O^R \delta W = \int_O^R 2\pi C dr = 2\pi C R$$

$$\therefore \quad C = \frac{W}{2\pi R}$$

Frictional torque on the elementory ring,

$$\delta T = 2\pi\mu p \text{ cosec } \alpha\, r^2 dr = 2\pi\mu \frac{C}{r} \text{ cosec } \alpha . r^2 dr$$
$$= 2\pi\mu C \text{ cosec } \alpha . r dr$$

Total frictional torque on the bearing,

$$T = \int_0^R \delta T = \int_0^R 2\pi\mu C \operatorname{cosec}\alpha . r dr$$

$$= 2\pi\mu C \operatorname{cosec}\alpha \left|\frac{r^2}{2}\right|_0^R$$

$$= \pi\mu C \operatorname{cosec}\alpha\ R^2$$

$$= \pi\mu \times \frac{W}{2\pi R} \times \operatorname{cosec}\alpha \cdot R^2$$

$$= \frac{1}{2}\mu W R \operatorname{cosec}\alpha$$

$$= \frac{1}{2}\mu W \ell$$

Example 5.4. **Vertical shaft of 200 mm diameter carries a load of 30 kN and runs at a speed of 140 rpm. The shaft is supported in a conical pivot bearing with cone angle 120° and coefficient of friction as 0.025. Find the power lost in friction for the following cases.**

(*i*) Uniform pressure

(*ii*) Uniform wear.

Solution :

$$D = 200 \text{ mm}$$
$$R = 100 \text{ mm} = 0.1 \text{ m}$$
$$W = 30 \text{ kN} = 30 \times 10^3 \text{ N.}$$
$$2\alpha = 120°$$
$$\alpha = 60°$$
$$\mu = 0.025$$
$$N = 140 \text{ rpm.}$$

(*i*) Uniform pressure

Total frictional torque on the bearing,

$$T = \frac{2}{3} \times \mu W R \operatorname{cosec}\alpha$$

$$= \frac{2}{3} \times 0.025 \times 30 \times 10^3 \times 0.1 \operatorname{cosec} 60°$$

$$= 57.7 \text{ Nm}$$

Power lost in friction

$$P = \frac{2\pi NT}{60 \times 1000} = \frac{2\pi \times 57.7 \times 140}{60 \times 1000} = \mathbf{0.846\ kW\ Ans.}$$

(*ii*) Uniform wear

Total frictional torque on the bearing

$$T = \frac{1}{2} \times \mu W R \operatorname{cosec}\alpha$$

$$= \frac{1}{2} \times 0.25 \times 30 \times 10^3 \times 0.1 \times \operatorname{cosec} 60° = 43.3 \text{ Nm}$$

The power lost in friction,

$$P = \frac{2\pi NT}{60 \times 1000} = \frac{2\pi \times 43.3 \times 140}{60 \times 1000} = \mathbf{0.635\ kW\ Ans.}$$

5.7.3. Trapezoidal (Truncated) Conical Pivot Bearing

A truncated conical pivot bearing is shown in Fig. 5.12.

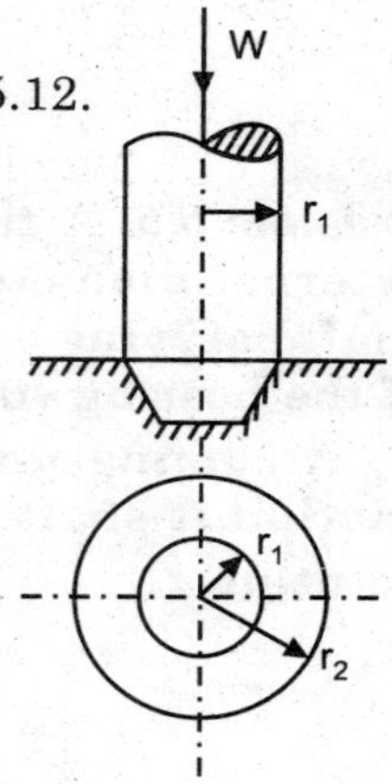

Fig. 5.12. Trapezoidal Pivot Bearing

Assume,

r_1 = External radius of bearing

r_2 = Internal radius of bearing

Area of bearing surface = $\pi[(r_1)^2 - (r_2)^2]$.

Normal pressure,

$$p = \frac{W}{A} = \frac{W}{\pi[(r_1)^2 - (r_2)^2]}$$

(a) Case of uniform pressure

Total frictional torque on the bearing

$$T = \int_{r_2}^{r_1} 2\pi\mu p \operatorname{cosec} \alpha r^2 dr$$

$$= 2\pi\mu p \operatorname{cosec}\alpha \left|\frac{r^3}{3}\right|_{r_2}^{r_1}$$

$$= 2\pi\mu p \operatorname{cosec}\alpha \left|\frac{(r_1)^3 - (r_2^3)}{3}\right|$$

$$= 2\pi\mu \frac{W}{\pi\left|(r_1)^2 - (r_2^2)\right|} \times \operatorname{cosec}\alpha \left|\frac{(r_1)^3 - (r_2)^3}{3}\right|$$

$$= \frac{2}{3}\mu W \operatorname{cosec}\alpha \left|\frac{(r_1)^3 - (r_2)^3}{(r_1)^2 - (r_2)^2}\right|$$

(b) Case of uniform wear

Total load on the bearing

$$W = \int_{r_2}^{r_1} 2\pi C dr = 2\pi C |r|_{r_2}^{r_1} = 2\pi C(r_1 - r_2)$$

$$\therefore \quad C = \frac{W}{2\pi(r_1 - r_2)}$$

Total frictional torque on the bearing

$$T = \int_{r_2}^{r_1} 2\pi\mu C \operatorname{cosec}\alpha . r dr = 2\pi\mu C \operatorname{cosec}\alpha \left|\frac{r^2}{2}\right|_{r_2}^{r_1}$$

$$= \pi\mu C \operatorname{cosec}\alpha[(r_1)^2 - (r_2)^2]$$

$$= \pi\mu \frac{W}{2\pi(r_1 - r_2)} \times \operatorname{cosec}\alpha[(r_1)^2 - (r_2)^2]$$

$$T = \frac{1}{2}\mu W(r_1 + r_2)\operatorname{cosec}\alpha$$

Assume mean radius of the bearing

$$R = \left(\frac{r_1 + r_2}{2}\right)$$

$$\therefore \quad T = \mu WR \operatorname{cosec} \alpha.$$

Example 5.5. **A truncated conical pivot bearing of cone angle 120°, external diameter twice the internal diameter supports a vertical shaft carrying a load of 20 kN. Find the internal and external radii of the bearing surface if normal pressure is not to exceed 0.3 N/mm².**

Assuming uniform pressure, calculate the power consumed in friction if shaft speed is 200 rpm and coefficient of friction 0.1.

Solution :

$$W = 20 \text{ kN} = 20 \times 10^3 \text{ N}.$$

$$2\alpha = 120°$$

$$\alpha = 60°$$

$$p = 0.3 \text{ N/mm}^2$$

$$\mu = 0.1$$

$$N = 200 \text{ rpm}$$

(*i*) Radii of bearing

$$r_1 = 2r_2$$

$$p = \frac{W}{A} = \frac{W}{\pi[(r_1)^2 - (r_2)^2]}$$

$$= \frac{W}{\pi[(2r_2)^2 - (r_2)^2]} = \frac{W}{3\pi r_2^2}$$

$$r_2 = \sqrt{\frac{W}{3\pi p}} = \sqrt{\frac{20\times10^3}{3\pi\times0.3}} = \mathbf{84 \text{ mm}}$$

$$r_1 = 2r_2 = \mathbf{168 \text{ mm Ans.}}$$

(*ii*) Friction power

The friction torque on the bearing

$$T = \frac{2}{3}\mu W \operatorname{cosec}\alpha\left[\frac{(r_1)^3 - (r_2)^3}{(r_1)^2 - (r_2)^2}\right]$$

$$= \frac{2}{3}\times0.1\times20\times10^3 \operatorname{cosec}60°\left[\frac{(168)^3 - (84)^3}{(168)^2 - (84)^2}\right]$$

$$= 301.76 \text{ Nm}$$

Power consumed by friction

$$P = \frac{2\pi NT}{60\times1000} = \frac{2\pi\times200\times301.76}{60\times1000} = \mathbf{6.322 \text{ kW Ans.}}$$

5.7.4. Flat Collar Bearings

The flat coller bearings are also called thrust bearings and may have single collar or multiple collars. The bearings are shown in Fig. 5.13.

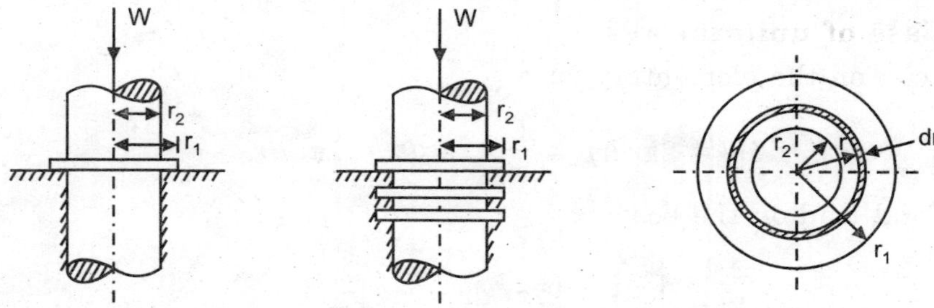

Fig. 5.13. Flat collar bearing

Assure r_1 = External radius of collar

r_2 = Internal radius of collar

Area of bearing, $A = \pi[(r_1)^2 - (r_2)^2]$ (Single collar bearing)

$= \pi n[(r_1)^2 - (r_2)^2]$ [Multi-collar bearing with number of collars n]

(*a*) Case of uniform pressure

The pressure,

$$p = \frac{W}{A} = \frac{W}{\pi[(r_1)^2 - (r_2)^2]}$$

The frictional torque on an elementary ring of radius r and thickness dr

$$\delta T = 2\pi\mu p r^2 dr$$

Total frictional torque,

$$T = \int_{r_2}^{r_1} \delta T = \int_{r_2}^{r_1} 2\pi\mu p r^2 dr$$

$$= 2\pi\mu p \left|\frac{r^3}{3}\right|_{r_2}^{r_1} = 2\pi\mu p\left[\frac{(r_1)^3 - (r_2)^3}{3}\right]$$

But $$p = \frac{W}{\pi[(r_1)^2 - (r_2^2)]}$$

$\therefore$ $$T = 2\pi\mu \frac{W}{\pi[(r_1)^2 - (r_2^2)]} \times \left[\frac{(r_1)^3 - (r_2)^3}{3}\right]$$

$$= \frac{2}{3}\mu W \frac{[(r_1)^3 - (r_2)^3]}{[(r_1)^2 - (r_2)^2]}$$

For multi-collar bearing

$$p = \frac{W}{n\pi[(r_1)^2 - (r_2)^2]}$$

$\therefore$ Total frictional torque

$$T = \frac{2}{3}\mu W\left[\frac{(r_1)^3 - (r_2)^3}{(r_1)^2 - (r_2)^2}\right]$$

(b) Case of uniform wear

Load on the elementary ring

$$\delta W = 2\pi r dr p = \frac{C}{r} \times 2\pi r dr = 2\pi C dr.$$

Total load on the bearing

$$W = \int_{r_2}^{r_1} \delta W = \int_{r_2}^{r_1} 2\pi C dr = 2\pi C |r|_{r_2}^{r_1}$$

$$= 2\pi C(r_1 - r_2)$$

$$\therefore \quad C = \frac{W}{2\pi(r_1 - r_2)}$$

Frictional torque on elementary ring

$$\delta T = \mu \delta W r = \mu \times 2\pi C dr.r = 2\pi\mu C r dr$$

Total frictional torque on the bearing

$$T = \int_{r_2}^{r_1} \delta T = \int_{r_2}^{r_1} 2\pi\mu C r dr = 2\pi\mu C \left|\frac{r^2}{2}\right|_{r_2}^{r_1}$$

$$= 2\pi\mu C \left|\frac{(r_1)^2 - (r_2)^2}{2}\right|$$

$$= \pi\mu C[(r_1)^2 - (r_2)^2]$$

$$= \pi\mu \frac{W}{2\pi(r_1 - r_2)}[(r_1)^2 - (r_2)^2]$$

$$= \frac{1}{2}\mu W(r_1 + r_2)$$

Example 5.6. **A thrust bearing used for marine propeller has 6 collars of 600 mm external diameter and 300 mm internal diameter. The shaft carriers a load of 100 kN and rotates at 90 rpm. Find the power consumed by friction for the following cases. Assure $\mu = 0.12$**

(*i*) Uniform pressure

(*ii*) Uniform wear

Solution :

$$d_1 = 600 \text{ mm}$$
$$r_1 = 300 \text{ mm} = 0.3 \text{ m}$$
$$d_2 = 300 \text{ mm}$$
$$r_2 = 150 \text{ mm} = 0.15 \text{ m}$$
$$n = 6$$
$$W = 100 \text{ kN} = 100 \times 10^3 \text{ N.}$$
$$\mu = 0.12$$
$$N = 90 \text{ rpm}$$

(i) Uniform pressure

The frictional torque

$$T = \frac{2}{3}\mu W\left[\frac{(r_1)^3-(r_2)^3}{(r_1)^2-(r_2)^2}\right]$$

$$= \frac{2}{3}\times 0.12\times 100\times 10^3\left[\frac{(300)^3-(150)^3}{(300)^2-(150)^2}\right]$$

$$= 2800 \times 10^3 \text{ Nmm} = 2800 \text{ Nm}$$

Power consumed in friction

$$P = \frac{2\pi NT}{60\times 1000} = \frac{2\pi\times 90\times 2500}{60\times 1000} = \textbf{26.4 kW Ans.}$$

(ii) Uniform wear

The frictional torque,

$$T = \frac{1}{2}\mu W(r_1+r_2) = \frac{1}{2}\times 0.12\times 100\times 10^3(300+150)$$

$$= 2700 \times 10^3 \text{ Nmm} = 2700 \text{ Nm}$$

The power consumed in friction

$$P = \frac{2\pi NT}{60\times 1000} = \frac{2\pi\times 90\times 2700}{60\times 1000} = \textbf{25.45 kW Ans.}$$

Example 5.7. **A thrust bearing of a propeller shaft consists of a number collars. The shaft is of 400 mm diameter and rotates at a speed of 90 rpm. The thrust on the shaft is 300 kN. If the intensity of pressure is to be 200 kN/m² and coefficient of fiction is 0.06, determine external diameter of the collars. The power lost in friction is not to exceed 48 kW.** ***[UPTU 2010-11]***

Solution :

$$d_2 = 400 \text{ mm}$$

$$r_2 = 200 \text{ mm} = 0.2 \text{ m}$$

$$W = 300 \text{ kN} = 300 \times 10^3 \text{ N}$$

$$N = 90 \text{ rpm}$$

$$\omega = \frac{2\pi N}{60} = \frac{2\pi\times 90}{60} = 9.42 \text{ rad/s.}$$

$$\mu = 0.06$$

$$p = 200 \text{ kN/m}^2 = \frac{200\times 1000}{1000\times 1000} \text{ N/mm}^2$$

$$= 0.2 \text{ N/mm}^2$$

$$P = 48 \text{ kW} = 48 \times 10^3 \text{ W.}$$

Power lost in friction,

$$P = T\omega$$

where T = Total frictional torque transmitted

$$\therefore \quad T = \frac{P}{\omega} = \frac{48\times 10^3}{9.42} = 5095.5 \text{ Nm.}$$

(i) External Diameter of the collar

Let external diameter of collar = $d_1 = 2r_1$

For uniform pressure,

$$T = \frac{2}{3}\mu W\left[\frac{(r_1)^3 - (r_2)^3}{(r_1)^2 - (r_2)^2}\right]$$

$$= \frac{2}{3}\mu W\left[\frac{(r_1)^2 + (r_2)^2 + r_1 r_2}{(r_1 + r_2)}\right]$$

$$5095.5 = \frac{2}{3}\times 0.06\times 300\times 10^3\left[\frac{(r_1)^2 + (0.2)^2 + 0.2\, r_1}{r_1 + 0.2}\right]$$

$$5095.5 \times 3(r_1 + 0.2) = 12 \times 10^3\ [(r_1)^2 + 0.2r_1 + 0.04]$$

Dividing both sides by 12×10^3

$$1.2738\ (r_1 + 0.2) = r_1^2 + 0.2r_1 + 0.04$$

$$r_1^2 - 1.0738\ r_1 - 0.21476 = 0$$

$$r_1 = \frac{1.0738 \pm \sqrt{(1.0738)^2 + 4(0.21476)}}{2}$$

$$= \frac{1.0738 \pm 1.41846}{2}$$

$= 1.246$ m [Taking +ve sign]

$= 1246$ mm

$d_1 = 2\ r_1 = 2 \times 1246 =$ **2492 mm Ans.**

(*ii*) Number of collars

The pressure, $$p = \frac{W}{n\pi[(r_1)^2 - (r_2)^2]}$$

The number of collars

$$n = \frac{W}{p\pi[(r_1)^2 - (r_2)^2]}$$

$$n = \frac{300\times 10^3}{0.2\times \pi[(1246)^2 - (200)^2]}$$

$$= \frac{300\times 10^3}{0.2\times \pi\times 151251.6} = 3.156 \text{ or } \mathbf{4\ Ans.}$$

QUESTION BANK

1. Explain the following:
   (*i*) Coefficient of friction
   (*ii*) Angle of friction
   (*iii*) Cone of friction
   (*iv*) Angle of repose
2. Explain the following:
   (*a*) Dry friction
   (*b*) Limiting friction
3. Explain law of static friction and dynamic friction.

4. Draw free body diagram of the following systems.
   (*i*) A body sliding down a rough inclined plane.
   (*ii*) A body being pulled up a rough inclined plane.
5. What is efficiency of an inclined plane.
6. Explain the following.
   (*a*) Friction in a journal bearing
   (*b*) Friction circle.
7. With the help of suitable sketches, explain various types of pivot and collar bearings.
8. Analyse a flat pivot bearing for the following conditions of operations.
   (*i*) Uniform pressure
   (*ii*) Uniform wear.
9. Find expressions for frictional torque in a conical pivot bearing for uniform pressure and uniform wear condition.
10. Explain the functioning of the following bearing.
    (*i*) Truncated conical pivot bearing
    (*ii*) Multi-collar bearing

NUMERICAL PROBLEMS

1. A conical pivot with angle of cone as 120°, supports a vertical shaft of diameter 200 mm. It is subjected to a load of 15 kN. The speed of shaft is 140 rpm. Takes $\mu = 0.05$. Calculate power lost in friction assuming.
   (*a*) Uniform pressure
   (*b*) Uniform wear [**Ans:** (*a*) 846 W (*b*) 634.8 W]
2. A plane collar thrust bearing has inner and outer diameter of 200 mm and 450 mm respectively. It is subjected to a vertical load of 40 kN. If $\mu = 0.25$ and speed of rotation 120 rpm, find the power lost in friction. The rate of wear is proportional to the pressure and rubbing velocity. [**Ans:** 20 kW].
3. A shaft has several integrated collars of 40 cm external diameter. The shaft diameter is 25 cm. If $\mu = 0.05$ and intensity of pressure is 350 kN/m^2, estimate
   (*i*) Power absorbed in friction if shaft runs at 105 rpm and carrier a load of 150 kN.
   (*ii*) No. of collars required [**Ans.** (*i*) 13.64 kW (*ii*) 6]

CHAPTER 6

BELT AND PULLEY DRIVE

6.1. INTRODUCTION

In industrial machines, power or motion has to be transmitted from one shaft (driver) to another shaft (driven). Various types of mechanism are used such as belt drive, rope drive, chain drive, gear boxes, etc. The belt and pulley drive is one of the most common method. The pulleys are mounted on the two shafts and an endless belt runs over the pulleys. The power of motion is transmitted due to friction between the pulleys and the belt. Therefore belt and pulley drives are not positive type of drives and there may be slippage. The velocity ratio is not always constant. There is line contact between the kinematic links and therefore belt and pulley drive belong to higher pairs of mechanism.

The belt and pulley drive is preferred over other mechanisms when the distance between the two shafts is large, power transmission is medium or small and very precise velocity ratio is not required.

The belt used may be flat or V-shape. A flat belt is used for small and medium power transmission. For more power transmission one or more V-belts are used which are inserted in a number of grooves in the pulley.

The flat belts are mostly made from leather, rubber, cotton or canvas whereas V-belts are made from stronger rubber impregnated fabric materials.

6.2. BELT DRIVE ARRANGEMENTS

Different belt drive arrangements are used for different applications. Some important arrangements are:

1. Open Belt Drive
2. Cross Belt Drive
3. Compound Belt Drive
4. Stepped Pulley Drive
5. Fast and Loose Pulley Drive
6. Belt Drive with Idler Pulleys.

6.2.1 Open Belt Drive

An open belt drive consists of a driver pulley (A) and driven pulley (B). The shafts are arranged in parallel. Both the pulleys rotate in the same direction, *i.e.*, clockwise direction as shown in Fig. 6.1.

The driver pulley pulls the belt and delivers to the driven pulley. The tension in the tight side of belt is more than tension in the slack side.

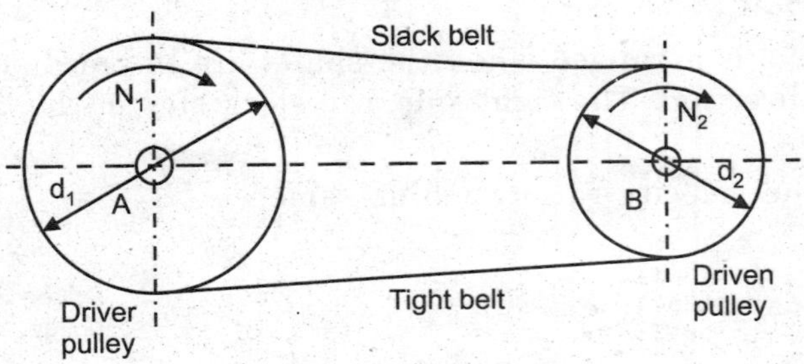

Fig. 6.1. Open belt drive

(*i*) Velocity Ratio

The velocity ratio (N_2/N_1) is the ratio of velocity of driven pulley to the velocity of driver pulley.

Let L_1 = Speed of driver pulley (rpm)

d_1 = Diameter of driver pulley (m)

N_2 = Speed of driven pulley (rpm)

d_2 = Diameter of driven pulley (m)

The length of belt passing over driver in one minute is equal to length of belt passing over driven pulley in one minute.

$$\pi d_1 N_1 = \pi d_2 N_2$$

Velocity ratio,

$$\frac{N_2}{N_1} = \frac{d_1}{d_2}$$

(*ii*) Effect of belt thickness (*t*)

For exact calculations, the mean diameter of rotation should be used. The mean diameter of rotation is the diameter of the pulley plugs belt thickness.

$$\therefore \quad \frac{N_2}{N_1} = \frac{d_1 + t}{d_2 + t}.$$

Example 6.1. **Calculate the speed of driven shaft carrying a pulley of 30 cm diameter. The engine diameter is 50 cm running at 200 rpm. What will be the speed if belt thickness is 10 mm?**

Solution : N_1 = 200 rpm

d_1 = 50 cm = 0.5 m

d_2 = 30 cm = 0.3 m

t = 10 mm = 0.01 m

(*i*) Without considering belt thickness.

$$N_2 = \frac{d_1}{d_2} \times N_1 = \frac{0.5}{0.3} \times 200 = \textbf{333.33 rpm Ans.}$$

(*ii*) Taking effect of belt thickness

$$N_2 = \frac{d_1 + t}{d_2 + t} \times N_2 = \frac{(0.5 + 0.01)}{(0.3 + 0.01)} \times 200 = \textbf{329.03 rpm Ans.}$$

6.2.2. Cross Belt Drive

Cross belt drive is arranged when the shafts are in parallel and rotate in the apposite direction. The tight side and slack side of the belt are shown in Fig. 6.2. The velocity ratio remains same *i.e.*, $\dfrac{N_2}{N_1} = \dfrac{d_1}{d_2}$

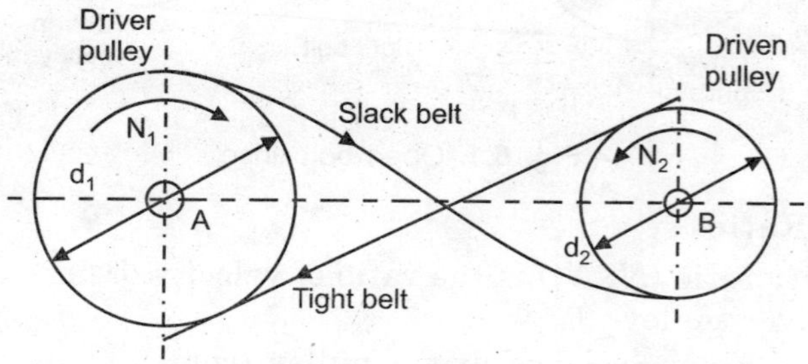

Fig. 6.2. Cross belt drive

6.2.3. Compound Belt Drive

When velocity ratios required are very large, an intermediate shaft is used to keep the diameter of pulleys within reasonable size. The intermediate shaft carriers two pulleys as shown in Fig. 6.3.

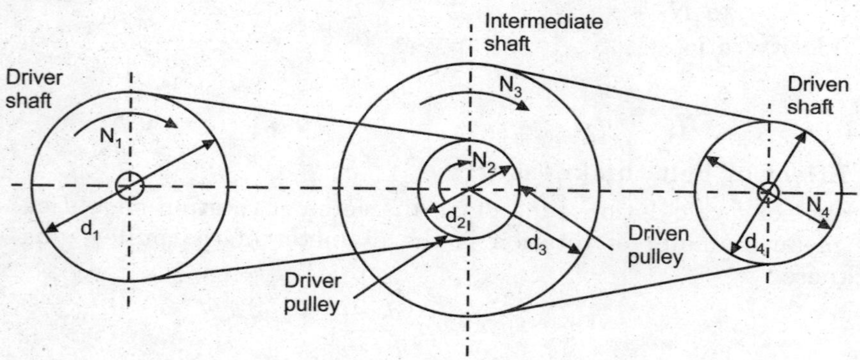

Fig. 6.3. Compound belt drive

$$\frac{N_2}{N_1} = \frac{d_1}{d_2}$$

$$\frac{N_4}{N_3} = \frac{d_3}{d_4}$$

$$\therefore \quad \frac{N_2}{N_1} \times \frac{N_4}{N_3} = \frac{d_1}{d_2} \times \frac{d_3}{d_4}$$

But $\quad N_2 = N_3$

$$\therefore \quad \frac{N_2}{N_1} \times \frac{N_4}{N_2} = \frac{d_1}{d_2} \times \frac{d_3}{d_4}$$

or velocity ratio, $\quad \dfrac{N_4}{N_1} = \dfrac{d_1 \times d_3}{d_2 \times d_4}$

Example 6.2. **An engine shaft running at 200 rpm is carrying a pulley of 80 cm diameter drives a pulley of 40 cm diameter on a line shaft. Another pulley of 100 cm mounted on the same line shaft drives a 20 cm diameter pulley mounted on a generator shaft. Determine the speed of the generator shaft.**

Solution : Speed ratio,

$$\frac{N_4}{N_1} = \frac{d_1 \times d_3}{d_2 \times d_4}$$

$$\therefore \quad N_4 = \frac{d_1 \times d_3}{d_2 \times d_4} \times N_1$$

$$= \frac{80 \times 100}{40 \times 20} \times 200 = 2000 \text{ rpm Ans.}$$

6.2.4. Stepped Pulley Drive

It is a variable speed drive where the speed of the drive shaft remains constant and speed of the driven shaft is varied by using pair of stepped pulleys. The diameters of the steps are selected, so that the same belt can be used and shifted onto different pair of stepped pulleys.

The stepped pulleys can be replaced by two cone pulleys which provide more number of speed ratios.

The following conditions must be satisfied in the design of stepped or cone pulleys.

1. The length of belt remains constant.
2. The distance between the two shaft remains constant.
3. The sum of radii of corresponding steps must be same.

$$\frac{N_2}{N_1} = \frac{d_1}{d_2} = \frac{r_1}{r_2}$$

$$\frac{N_4}{N_3} = \frac{r_3}{r_4}$$

$$\therefore \quad \frac{N_6}{N_5} = \frac{r_5}{r_6}$$

$$r_1 + r_2 = r_3 + r_4 = r_5 + r_6$$

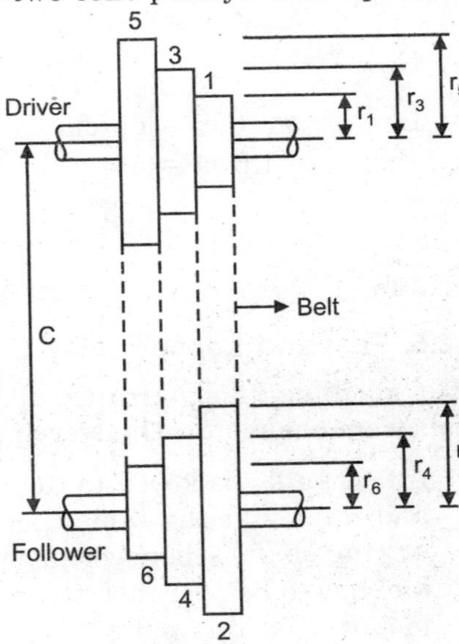

Fig. 6.4. Stepped Pulley Drive.

The length of belt for open drive,

$$L = \pi(r_1 + r_2) + \frac{(r_1 - r_2)^2}{C} + 2C.$$

Example 6.3. **Design a crossed belt stepped pulley drive to achieve speeds of 60 rpm, 80 rpm and 100 rpm from a constant speed motor at 160 rpm. The distance between parallel shafts should be 180 cm and radius of smallest pulley should be 15 cm. Neglect belt thickness and slip.**

Solution : Refer Fig. 6.4.

$$N_1 = N_3 = N_5 = 160 \text{ rpm.}$$

$N_2 = 60$ rpm, $N_4 = 80$ rpm, $N_6 = 100$ rpm.

$C = 180$ cm

$r_1 = 15$ cm

$$\frac{N_2}{N_1} = \frac{d_1}{d_2} = \frac{r_1}{r_2}$$

$$\therefore \quad r_2 = r_1 \times \frac{N_1}{N_2} = 15 \times \frac{160}{60} = \textbf{40 cm Ans.}$$

Similary, $$r_4 = r_3 \times \frac{N_3}{N_4} = r_3 \times \frac{160}{80} = 2r_3$$

Now, $r_1 + r_2 = r_3 + r_4 = r_5 + r_6$

$r_1 + r_2 = 15 + 40 = 55$ cm

$r_3 + r_4 = 55$ cm

$r_3 + 2r_3 = 55$ cm

$$r_3 = \frac{55}{3} = \textbf{18.33 cm Ans.}$$

$r_4 = 2 \times 18.33 =$ **36.66 cm Ans.**

Now $$r_6 = r_5 \times \frac{N_5}{N_6} = r_5 \times \frac{160}{100} = 1.6\, r_5$$

But $r_5 + r_6 = 55$ cm

$r_5 + 1.6\, r_5 = 55$

$$\therefore \quad r_5 = \frac{55}{2.6} = \textbf{21.15 cm Ans.}$$

and $r_6 = 1.6 \times 21.15 =$ **33.85 cm Ans.**

6.2.5. Fast and Loose Pulley Drive

In a machine shop, arrangement can be made to drive several machine tools from one motor. On every machine tool shaft two pulleys are mounted.

1. A fast pulley is keyed to the shaft and rotates at the same speed as driving shaft. Whenever, machine tool has to be operated, the belt is mounted on this pulley.
2. A loose pulley is not keyed to the shaft and is free to rotate on the machine tool shaft. Whenever, the machine has to be stopped (while the motor is running), the belt is pushed onto loose pulley from the fast pulley with the help of a forked sliding bar. The belt can be pushed back onto fast pulley whenever machine tool has to be operated.

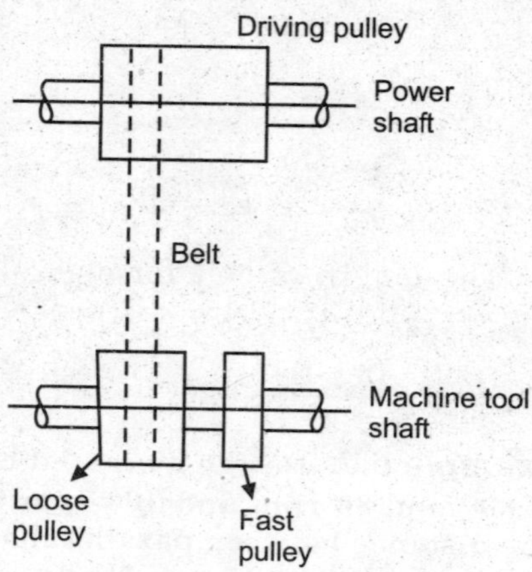

Fig. 6.5. Fast and loose pulley drive.

6.2.6 Belt Drive with Idler Pulleys

Idler pulleys are used in open belt dive to acheive the following functions:

1. The power capacity of a belt drive can be restored or enhanced by increasing the belt tension and angle of contact. This can be done by applying pressure force on the belt with the help of an idler pulley as shown in Fig. 6.6.

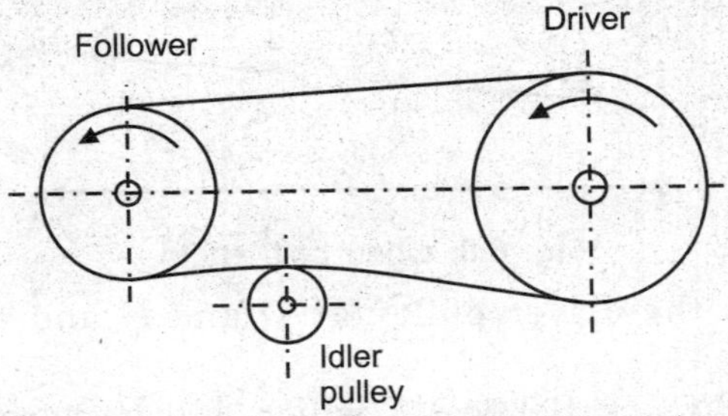

Fig. 6.6. Idler Pulley

2. The life of belt is also improved.
3. The idler pulleys can help to transmit power from one shaft to several shafts simultaneously as shown in Fig. 6.7. The idler pulleys are loose pulleys.

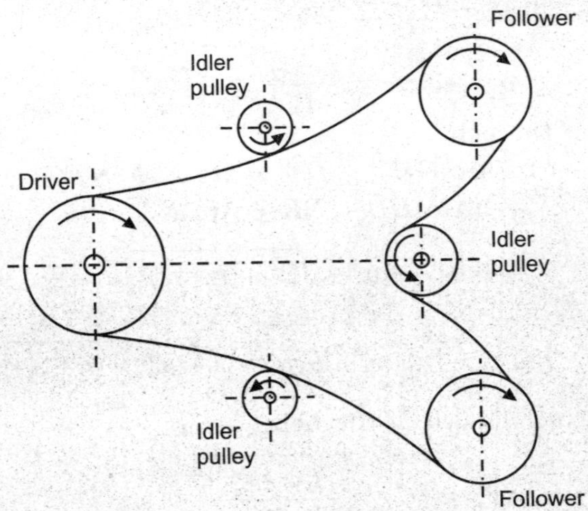

Fig. 6.7. Idler Pulleys Arrangement

6.3. LENGTH OF OPEN BELT DRIVE

The total length of the belt can be divided into three parts:

1. Belt length embracing the driver pulley (DHG)
2. Belt length embracing the driven pulley (EJF)
3. Belt length not in contact with either pulley (DE+FG)

An open belt drive is shown in Fig. 6.8.

Let $C = AB$ = Distance between centres of two shafts.

$r_1 = AD = AG$ = Radius of driver pulley

$r_2 = BE = BF$ = Radius of driven pulley.

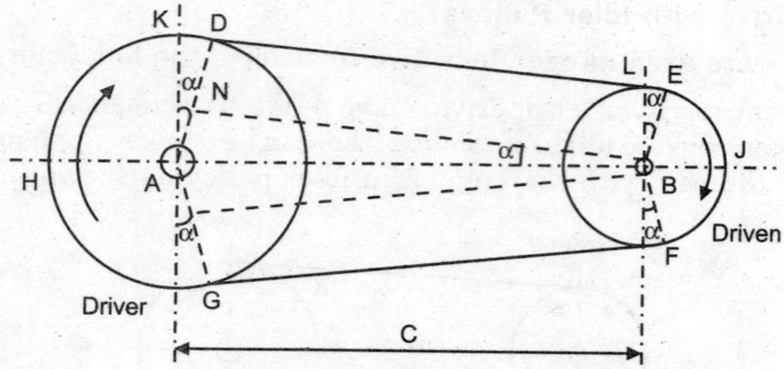

Fig. 6.8. Open belt length

The belt leaves the driver pulley at D and G and driver pulley at E and F.

From B draw BN parallel to DE which will be perpendicular to AD.

Let $\angle KAD = \angle LBE = \alpha$

$$AN = AD - DN = r_1 - r_2$$

$$BN = \sqrt{AB^2 - AN^2} = \sqrt{C^2 - (r_1 - r_2)^2}$$

From ΔABN, $\sin\alpha = \dfrac{AN}{AB} = \dfrac{r_1 - r_2}{C}$

or $$\alpha = \sin\alpha = \frac{r_1 - r_2}{C} \qquad [\because \alpha \text{ is very small}]$$

The length of the belt,

$$L = \text{Arc } GHD + DE + \text{Arc } EJF + FG$$

$$= 2[\text{Arc } HD + DE + \text{Arc } EJ] \qquad [\because DE = FG]$$

$$= 2\left[r_1\left(\frac{\pi}{2} + \alpha\right) + \sqrt{C^2 - (r_1 - r_2)^2} + r_2\left(\frac{\pi}{2} - \alpha\right)\right]$$

$$= 2\left[\frac{\pi}{2}(r_1 + r_2) + \alpha(r_1 - r_2) + \sqrt{C^2 - (r_1 - r_2)^2}\right]$$

This is the exact length of the belt.

Note: $$\sqrt{C^2 - (r_1 - r_2)^2} = C\left[\sqrt{1 - \frac{(r_1 - r_2)^2}{C}}\right]$$

$$= C\left[1 - \frac{1}{2}\left(\frac{r_1 - r_2}{C}\right)^2 + \ldots\right] \qquad \text{[By Binomial expansion]}$$

$$= C\left[1 - \frac{(r_1 - r_2)^2}{2C^2}\right] \qquad \text{[Neglecting smaller values]}$$

$\therefore$ $$L = 2\left\{r_1\left(\frac{\pi}{2} + \alpha\right) + C\left[1 - \frac{(r_1 - r_2)^2}{2C^2}\right] + r_2\left(\frac{\pi}{2} - \alpha\right)\right\}$$

$$= 2\left\{\frac{\pi}{2}(r_1 + r_2) + \alpha(r_1 - r_2) + \left[C - \frac{(r_1 - r_2)^2}{C^2}\right]\right\}$$

$$= \pi(r_1 + r_2) + 2\alpha(r_1 + r_2) + 2C - \frac{(r_1 - r_2)^2}{C}$$

But $$\alpha = \frac{r_1 - r_2}{C}$$

$$\therefore \quad L = \pi(r_1 + r_2) + \frac{2}{C}(r_1 + r_2)^2 + 2C - \frac{(r_1 - r_2)^2}{C}$$

$$L = \pi(r_1 + r_2) + \frac{(r_1 - r_2)^2}{C} + 2C$$

This is approximate length of the belt.

Example 6.4. **Two parallel shafts 6 m apart are connected by an open belt running over pulleys of 60 cm and 40 cm diameter. Determine the exact and approximate length of the belt.**

Solution : Data given $C = 6 \text{ m} = 600 \text{ cm}$

$$r_1 = \frac{60}{2} = 30 \text{ cm}$$

$$r_2 = \frac{40}{2} = 20 \text{ cm.}$$

(*i*) Exact length

$$L = 2\left[\frac{\pi}{2}(r_1 + r_2) + \alpha(r_1 - r_2) + \sqrt{C^2 - (r_1 - r_2)^2}\right]$$

$$\alpha \approx \sin \alpha = \frac{r_1 - r_2}{C}$$

$$\alpha = \sin^{-1}\left(\frac{30 - 20}{600}\right) = 0.955° = 0.955 \times \frac{\pi}{180} = 0.01606 \text{ rad.}$$

$$\therefore \quad L = 2\left[\frac{\pi}{2}(30 + 20) + 0.01666(30 - 20) + \sqrt{600^2 - (30 - 20)^2}\right]$$

$$= 2[25\pi + 0.1666 + 600]$$

$$= \mathbf{1357.3 \text{ cm Ans.}}$$

(*ii*) Approximate length

$$L = \pi(r_1 + r_2) + \frac{(r_1 - r_2)}{C} + 2C$$

$$= \pi(30 + 20) + \frac{(30 - 20)}{600} + 2 \times 600 = \mathbf{1357.16 \text{ cm Ans.}}$$

6.4. LENGTH OF CROSS BELT DRIVE

A cross belt drive is shown in Fig 6.9. with the following data.

r_1 = Radius of driver pulley
r_2 = Radius of driven pulley
C = Distance between centres of the shafts
L = Total length of cross belt

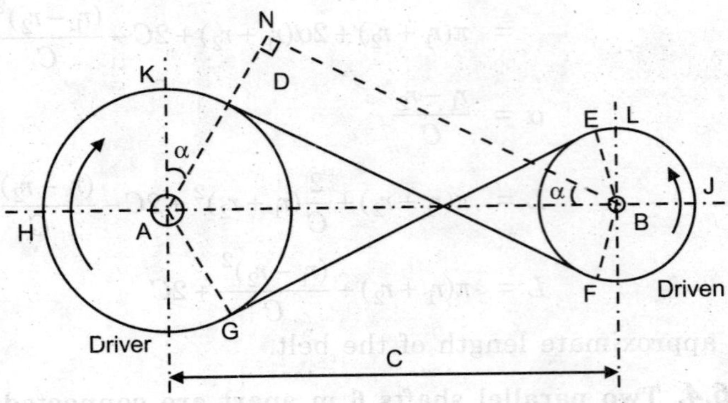

Fig. 6.9. Cross belt length

The belt leaves the driver at point D and G and driven pulley at E and F.

From B draw BN parallel to DF and perpendicular to AD

$$\angle ADF = \angle ANB = 90^\circ$$

Let $\angle ABN = \alpha$

$\therefore$ $\angle KAD = \angle EBL = \alpha$

Now $AN = AD + DN$

$$= r_1 + BF = r_1 + r_2$$

In ΔABN, $BN = \sqrt{AB^2 - AN^2} = \sqrt{C^2 - (r_1 + r_2)^2}$

$$\sin\alpha = \frac{AN}{AB} = \frac{r_1 + r_2}{C}$$

Now $\sin\alpha = \alpha$ being very small

$$\alpha = \frac{r_1 + r_2}{C}$$

Total length of belt,

$$L = \text{Arc } GHD + DF + \text{Arc } FJE + EG$$

$$= [\text{Arc } HD + DF + \text{Arc } JE] \qquad [\because DF = EG]$$

$$= 2\left[r_1\left(\frac{\pi}{2}+\alpha\right) + DF + r_2\left(\frac{\pi}{2}+\alpha\right)\right]$$

$$= 2\left[r_1\left(\frac{\pi}{2}+\alpha\right) + \sqrt{C^2 - (r_1 + r_2)^2} + r_2\left(\frac{\pi}{2}+\alpha\right)\right]$$

$$L = 2\left[\frac{\pi}{2}(r_1 + r_2) + \alpha(r_1 + r_2) + \sqrt{C^2 - (r_1 + r_2)^2}\right]$$

This is the exact length.

Now $\sqrt{C^2 - (r_1 + r_2)^2} = C\left[\sqrt{1 - \left(\frac{r_1 + r_2}{C}\right)^2}\right]$

$$= C\left[1 - \frac{1}{2}\left(\frac{r_1 + r_2}{C}\right)^2 + \cdots\right] \quad \text{[using Binomial expansion]}$$

$$\sqrt{C^2 - (r_1 + r_2)^2} = C\left[1 - \frac{(r_1 + r_2)^2}{2C^2}\right] \quad \text{[Neglecting smaller value]}$$

$$= C - \frac{(r_1 + r_2)^2}{2C}$$

$$\therefore \quad L = \pi(r_1 + r_2) + 2\left(\frac{r_1 + r_2}{C}\right)(r_1 + r_2) + 2C - \frac{(r_1 + r_2)^2}{C}$$

$$= \pi(r_1 + r_2) + \frac{2}{C}(r_1 + r_2)^2 + 2C - \frac{(r_1 + r_2)^2}{C}$$

$$= \pi(r_1 + r_2) + \frac{(r_1 + r_2)^2}{C} + 2C.$$

This is the approximate length of the belt.

Example 6.5. **Two parallel shafts 6 m apart are carrying pulleys of diameter 60 cm and 40 cm. Determine the exact and approximate length of a cross belt.**

Solution : $C = 60 \text{ m} = 600 \text{ cm}$

$$r_1 = \frac{60}{2} = 30 \text{ cm}$$

$$r_2 = \frac{40}{2} = 20 \text{ cm}$$

$$\alpha = \sin^{-1}\frac{r_1 + r_2}{C} = \sin^{-1}\frac{30 + 20}{600} = 4.78°$$

$$= 4.78 \times \frac{\pi}{180} = 0.08343 \text{ rad}$$

The exact length of cross belt

$$L = 2\left[\frac{\pi}{2}(r_1 + r_2) + \alpha(r_1 + r_2) + \sqrt{C^2 - (r_1 + r_2)^2}\right]$$

$$= 2\left[\frac{\pi}{2}(30 + 20) + 0.08343(30 + 20) + \sqrt{600^2 - (30^2 + 20^2)}\right]$$

$$= 2[25\pi + 4.1715 + 597.913] = \textbf{1361.24 cm Ans.}$$

The approximate length of belt

$$L = \pi(r_1 + r_2) + \frac{(r_1 + r_2)^2}{C} + 2C$$

$$= \pi(30 + 20) + \frac{(30 + 20)^2}{600} + 2 \times 600 = \textbf{1361.31 cm Ans.}$$

6.5. RATIO OF BELT TENSION FOR FLAT BELT DRIVE

Power is transmitted from one shaft to another by friction between the belt and the pulleys on which it passes. The driving pulley pulls the belt from one side and delivers the same to the other side. The belt tension T_1 on tight side is move then tension T_2 on slack side.

An open belt drive system is shown in Fig. 6.10.

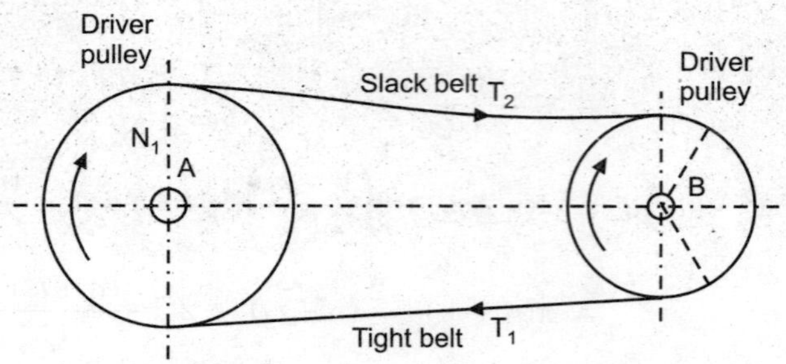

Fig. 6.10. Open belt drive with tension

Consider the driven pulley *B*.

T_1 = Tension in the belt on tight side.
T_2 = Tension in the belt on slack side
r = Radius of the pulley
μ = Coefficient of friction between belt and pulley
θ = Angle of contact between belt and pulley

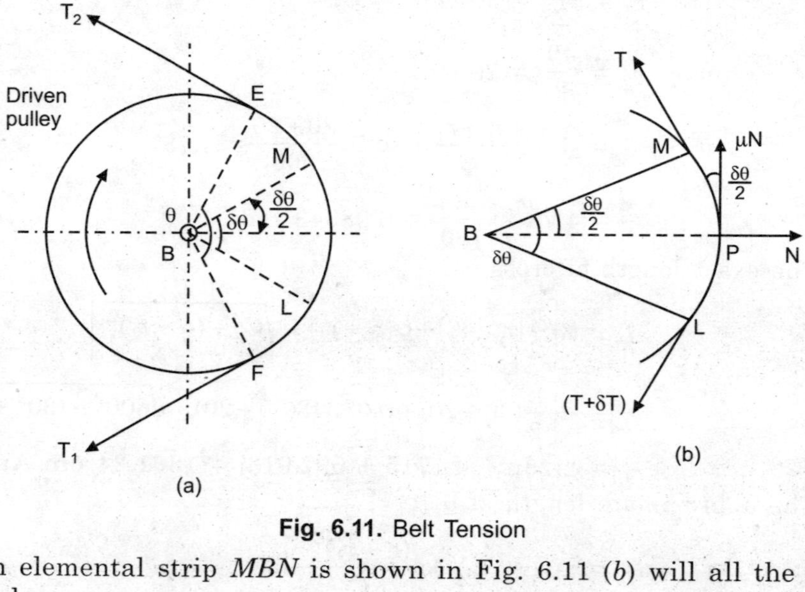

Fig. 6.11. Belt Tension

An elemental strip *MBN* is shown in Fig. 6.11 (*b*) will all the forces marked.

The following forces keep the elemental strip in equilibrium

1. Tension *T* acting tangentially at point *M*.
2. Tension $(T + \delta T)$ acting tangentially at point *L*
3. Normal reaction *N* at *P* acting radially outwards
4. Force of friction μN acting at right angle to *N* and in a direction opposing the motion.

$$\angle PBM = \frac{\delta\theta}{2} \text{ (rad)}$$

Resolving the forces in the horizontal direction

$$N = T \sin\frac{\delta\theta}{2} + (T + \delta T)\sin\frac{\delta\theta}{2}.$$

Angle $\delta\theta$ is very small

$$\therefore \quad \sin\frac{\delta\theta}{2} \approx \frac{\delta\theta}{2}$$

$$\therefore \quad N = T \times \frac{\delta\theta}{2} + (T + \delta P)\frac{\delta\theta}{2}$$

$$= T \times \frac{\delta\theta}{2} + T \times \frac{\delta\theta}{2} + \delta T \times \frac{\delta\theta}{2}$$

$$= 2T \times \frac{\delta\theta}{2} + \frac{\delta T \times \delta\theta}{2}$$

$$N = T.\delta\theta \quad \ldots(i) \quad \left[\text{Neglecting } \frac{\delta T \times \delta\theta}{2} \text{ being very small}\right]$$

Resolving the forces in the vertical direction

$$\mu N = (T + \delta T)\cos\frac{\delta\theta}{2} - T\cos\frac{\delta\theta}{2}$$

$$= T + \delta T - T \qquad \left[\cos\frac{\delta\theta}{2} = 1 \text{ for small } \delta\theta\right]$$

$$= \delta T$$

$$\therefore \quad N = \frac{\delta T}{\mu} \quad \ldots(ii)$$

Equating equations (*i*) and (*ii*)

$$T\delta\theta = \frac{\delta T}{\mu}$$

$$\frac{\delta T}{T} = \mu\delta\theta.$$

Integrating the equation between limits T_2 and T_1

$$\int_{T_2}^{\pi} \frac{\delta T}{T} = \int \mu\, \delta\theta$$

$$\ln\frac{T_1}{T_2} = \mu\theta$$

$$\frac{T_1}{T_2} = e^{\mu\theta}$$

This is the ratio of tensions in flat belt drive.

Notes: 1. For open belt drive,

Angle of contact $\theta = (180° - 2\alpha)$

where $\sin\alpha = \dfrac{r_1 - r_2}{C}$

2. For cross belt drive

Angle of contact $\theta = (180° + 2\alpha)$

where $\sin \alpha = \dfrac{r_1 + r_2}{C}$.

3. Power transmitted by belt,

$$P = \frac{(T_1 - T_2) \times v}{1000} \text{ kW}$$

where effecting driving force $= (T_1 - T_2)$

v = velocity of the belt (m/s)

Torque exerted on driven pulley $= (T_1 - T_2)\, r_2$

Torque exerted by driving pulley $= (T_1 - T_2)\, r_1$

Example 6.6. **Find the power transmitted by a belt running over a pulley of 600 mm diameter at 200 rpm. The coefficient of friction between the belt and pulley is 0.25, angle of lap 160° and maximum tension in the belt is 2.5 kN.**

Solution : $r = \dfrac{600}{2} = 300 \text{ mm} = 0.3 \text{ m}, \quad \mu = 0.25, \ \theta = 160°$

$$T_1 = 2.5 \text{ kN}$$

$$\frac{T_1}{T_2} = e^{\mu\theta}$$

$$\theta = 160° \times \frac{\pi}{180} = 2.79 \text{ rad.}$$

$$\frac{T_1}{T_2} = e^{0.25 \times 2.79} = 2.01$$

$$\therefore \quad T_2 = \frac{T_1}{2.01} = \frac{2.5}{2.01} = 1.244 \text{ kN}$$

$$P = \frac{2\pi N(T_1 - T_2)r}{60} = \frac{2\pi \times 200(2.5 - 1.244) \times 0.3}{60}$$

$= \textbf{7.89 kW Ans.}$

Example 6.7. **An open belt drive connects two parallel shafts 4 m apart carrying pulleys of 120 cm and 50 cm diameters. The driver pulley runs at 200 rpm. The coefficient of friction is 0.3 and maximum tension in the belt is 1855 N.**

Calculate ***(i)*** **The power transmitted**

(ii) **Torque on each shaft.**

Solution : Driver pulley, $d_1 = 120 \text{ cm} = 1.2 \text{ m}$

$r_1 = 0.6 \text{ m}$

Driven pulley, $d_2 = 50 \text{ cm} = 0.5 \text{ m}$

$r_2 = 0.25 \text{ m}$

$\mu = 0.3$

$T_1 = 1855 \text{ N.}$

$N_1 = 200 \text{ rpm}$

Velocity of belt, $v = \dfrac{\pi d_1 N_1}{60} = \dfrac{\pi \times 1.2 \times 200}{60} = 12.56 \text{ m/s}$

Angle of contact, $\theta = 180 - 2\alpha$

where $\sin\alpha = \dfrac{r_1 - r_2}{C} = \dfrac{0.6 - 0.25}{4} = 0.0875$

$\alpha = 5.02°$

$\therefore$ $\theta = 180 - 2 \times 5.02 = 169.96°$

$= 169.96 \times \dfrac{\pi}{180} = 2.967$ rad.

Now, $\dfrac{T_1}{T_2} = e^{\mu\theta} = e^{0.3 \times 2.967} = 2.435$

$\therefore$ $T_2 = \dfrac{1855}{2.435} = 761.8$ N

(*i*) Power transmitted,

$$P = \frac{(T_1 - T_2)v}{1000} = \frac{(1855 - 761.8) \times 12.56}{1000}$$

$= $ **1373 kW Ans.**

(*ii*) Torque exerted by driving shaft

$\tau_1 = (T_1 - T_2) \times r_1$

$= (1855 - 761.8) \times 0.6 = $ **856 Nm Ans.**

Torque exerted on driven shaft

$\tau_2 = (T_1 - T_2) \times r_2$

$= (1855 - 761.8) \times 0.25 = $ **273.4 Nm Ans.**

6.6. CENTRIFUGAL TENSION

The belt has mass *m* and is rotated in a circular path of radius *r* of the rim of the pulley with a uniform velocity *v*. Therefore a centrifugal force acts radially outwards with a magnitude,

$$F_c = \frac{mv^2}{r}$$

This force can cause the following effects on the belt.

1. A centrifugal tension is produced in the belt in addition to tension due to power transmission.
2. Centrifugal force tends to left the belt from the rim of the pulley. The normal reaction (*N*) and frictional resistance ($f = \mu N$) are reduced.
3. The driving power capacity (*P*) is reduced.
4. The centrifugal force is negligible at very low speeds

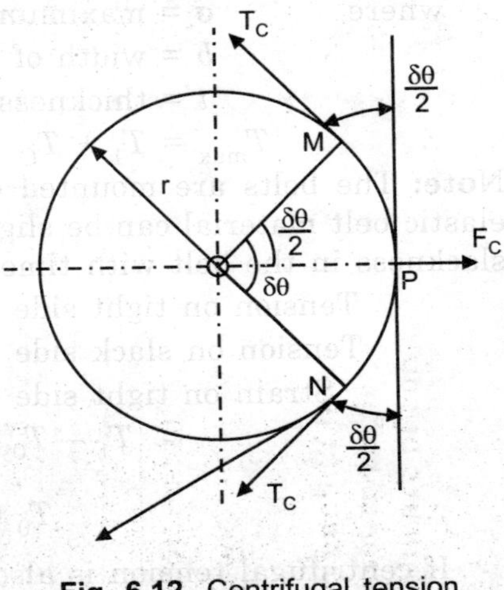

Fig. 6.12. Centrifugal tension

Consider an elementary length of belt MN substending an angle $\delta\theta$ at the centre of the shaft as shown in Fig. 6.12.

Let m = mass of belt per unit length (kg/m)

M = total mass of elementry length of belt (kg)

v = linear velocity of belt (m/s)

r = radius of the pulley (m)

T_C = Centrifugal tension at sections M and N acting tangentially (N/m²)

F_C = Centrifugal force acting radially outwards (N)

$$\text{Arc } MN = r.\delta\theta$$

$$M = m \times r\delta\theta$$

$$F_C = M\frac{v^2}{r} = mr\delta\theta\frac{v^2}{r}.$$

For quilibrium of elementary length of belt, resolving forces horiontally,

$$T_C \sin\frac{\delta\theta}{2} + T_C \sin\frac{\delta\theta}{2} = F_C$$

$$2T_C \sin\frac{\delta\theta}{2} = mr\delta\theta\frac{v^2}{r}$$

For very small value of $\delta\theta$, $\sin\frac{\delta\theta}{2} = \frac{\delta\theta}{2}$

$$\therefore \quad 2T_C \frac{\delta\theta}{2} = mr\delta\theta\frac{v^2}{r}$$

$$\therefore \quad T_C = mv^2.$$

Total tension of tight side of belt = $T_1 + T_C$

Total tension on slack side of belt = $T_2 + T_C$

The maximum tension in the belt is the maximum safe stress in the belt.

$$T_{max} = \sigma \times b \times t$$

where σ = maximum safe stress is the belt.

b = width of the belt

t = thickness of the belt

$$\therefore \quad T_{max} = T_1 + T_C$$

Note: The belts are mounted on the pulleys with some tension so that elastic belt material can be slightly stretched and can compensate for the slackness in the belt with time. It is called **initial tension** (T_0).

Tension on tight side = $T_1 - T_0$

Tension on slack side = $T_0 - T_2$

Strain on tight side = Strain on slack side

$$T_1 - T_0 = T_0 - T_2$$

$$\therefore \quad T_0 = \frac{T_1 + T_2}{2}$$

If centrifugal tension is also considered

$$T_0 = \frac{T_1 + T_2 + 2T_C}{2}$$

Example 6.7. **Determine the maximum power that can be transmitted by a belt of width 20 cm and thickness of 8 mm and density of material 1 g/cm³. The maximum permissible stress in the belt should not exceed 250 N/cm². The contact angle is 165° and the speed of belt is 1700 m/min. Take μ = 0.25.**

Solution :

$$\theta = 165° = 165 \times \frac{\pi}{180} = 2.88 \text{ rad}$$

$$v = 1700 \text{ m/min} = \frac{1700}{60} \; 28.33 \text{ m/s}$$

$$b = 20 \text{ cm}$$

$$t = 8 \text{ mm} = 0.8 \text{ cm}$$

$$\rho = 1 \text{ g/cm}^3 = \frac{1}{1000} \text{ kg/cm}^3$$

$$\sigma_{max} = 250 \text{ N/cm}^2$$

$$\mu = 0.25$$

$$T_{max} = \sigma_{max} \times b \times t$$

$$= 250 \times 20 \times 0.8 = 4000 \text{ N}$$

$$T_C = m \times v^2$$

$$= (\rho \times b \times t \times L)\, v^2$$

$$= \frac{1}{1000} \times 20 \times 0.8 \times 100 \times 28.33^2$$

[Assume L = 1 m = 100 cm]

$$= 1284 \text{ N}$$

$$T_1 = T_{max} - T_C = 4000 - 1284 = 2716 \text{ N}$$

$$\frac{T_1}{T_2} = e^{\mu\theta} = e^{0.25 \times 2.88} = 2.056$$

$$T_2 = \frac{T_1}{2.056} = \frac{2716}{2.056} = 1321 \text{ N.}$$

$$P = \frac{(T_1 - T_2)v}{1000} = \frac{(2716 - 1321)}{1000} \times 28.83$$

$$= \mathbf{39.52 \text{ kW Ans.}}$$

6.7. CONDITION FOR MAXIMUM POWER TRANSMISSION

The power transmitted by a belt,

$$P = (T_1 - T_2) \times v \; [W]$$

where T_1 = Tension on tight side (N)

T_2 = Tension on slack side (N)

v = Linear velocity of belt (m/s)

Also $$\frac{T_1}{T_2} = e^{\mu\theta}$$

or $$T_2 = \frac{T_1}{e^{\mu\theta}}$$

$$\therefore \quad P = \left(T_1 - \frac{T_1}{e^{\mu\theta}}\right)v = T_1\left(1 - \frac{1}{e^{\mu\theta}}\right)v$$

$$= T_1 \times k \times v$$

where $$k = \left(1 - \frac{1}{e^{\mu\theta}}\right)$$

$$T_{max} = T_1 + T_C$$

or $$T_1 = T_{max} - T_C$$

$$\therefore \quad P = k(T_{max} - T_C) \times v$$

$$= k(T_{max} - mv^2) \times v \qquad [\because T_C = m \times v^2]$$

$$= k[T_{max}\, v - mv^3]$$

For maximum power, $\frac{dP}{dv} = 0$

$$\therefore \quad \frac{dP}{dv} = [T_{max} - 3mv^2] = 0$$

$$k \neq 0$$

$$\therefore \quad T_{max} - 3mv^2 = 0$$

$$T_{max} = 3mv^2$$

or $$v = \sqrt{\frac{T_{max}}{3m}}$$

Now $$T_C = mv^2$$

$$\therefore \quad T_{max} = 3T_C$$

But $$T_{max} = T_1 + T_C$$

$$= T_1 + \frac{T_{max}}{3}$$

$$\therefore \quad T_1 = T_{max} - \frac{T_{max}}{3} = \frac{2}{3}T_{max}$$

$$P_{max} = \frac{2}{3}T_{max}\left(1 - \frac{1}{e^{\mu\theta}}\right)\sqrt{\frac{T_{max}}{3m}}$$

Condition for maximum power transmission.

1. $$v = \sqrt{\frac{T_{max}}{3m}}$$

2. $$T_C = \frac{T_{max}}{3}$$

3. $$T_1 = \frac{2}{3}T_{max}.$$

Example 6.8. **Determine the maximum power that can be transmitted through a flat belt having the following data:**

Cross-section of the belt = 300 mm × 12 mm
Ratio of friction tension = 2.4
Maximum permissible tension = 2 MPa
Mass density of belt material = 1.1 × 10^{-3} g/mm^3 [UPTU 2010-11]

Solution: $\rho = 1.1 \times 10^{-3}$ g/mm^3

$$= \frac{1.1}{1000} \text{ kg/cm}^3$$

$$\sigma_{max} = 2 \text{ MPa} = 2 \times 10^6 \text{ N/m}^2 = 200 \text{ N/cm}^2$$

$$b = 300 \text{ mm} = 30 \text{ cm}$$

$$t = 12 \text{ mm} = 1.2 \text{ cm}$$

$$\frac{T_1}{T_2} = 2.4$$

$$m = \text{Density} \times \text{vol of belt per meter length}$$

$$= \frac{1.1}{1000} \times 30\ 1.2 \times 100 = 3.96 \text{ kg}$$

$$T_{max} = \sigma_{max} \times b \times t$$

$$= 200 \times 30 \times 1.2 = 7200 \text{ N.}$$

For condition of maximum power transmission,

$$v = \sqrt{\frac{T_{max}}{3m}} = \sqrt{\frac{7200}{3 \times 3.96}} = 24.6 \text{ m/s.}$$

$$T_C = \frac{T_{max}}{3} = \frac{7200}{3} = 2400 \text{ N}$$

$$T_{max} = T_1 + T_C$$

$$\therefore \quad T_1 = T_{max} - T_C = 7200 - 2400 = 4800\ N$$

$$\frac{T_1}{T_2} = 2.4$$

$$T_2 = \frac{T_1}{2.4} = \frac{4800}{2.4} = 2000 \text{ N}$$

$$P_{max} = \frac{(T_1 - T_2) \times v}{1000} = \frac{(4800 - 2000) \times 24.6}{1000}$$

$$= \mathbf{68.88 \text{ kW Ans.}}$$

6.8. V-BELT DRIVE

V-belt drive is mostly used in the industry where large amount of power is transmitted and distance between shafts is very small. The V-belts are mounted inside suitable grooves made in the rim of the pulley. The power is transmitted due to wedging action between the belt and V-groove in the pulley.

The V-belt drive has many advantages.

1. The drive is positive with negligible slip.
2. The installation is very compact.
3. The running is very smooth, and quiet.
4. The drive is very durable.
5. High velocity ratios and power transmission are possible.
6. The V-belt drives can be operated in horizontal, vertical or inclined position. The same belt can be operated in either direction.

A V-belt with a grooved pulley is shown in Fig. 6.13 alongwith various forces acting.

2α = angle of groove

μ = coefficient of friction between belt and pulley.

N = Normal reaction between belt and sides of grooved pulley.

The total forces acting vertically,

$$R = N \sin \alpha + N \sin \alpha$$
$$= 2N \sin \alpha.$$
$$N = \frac{R}{2\sin\alpha} = \frac{R}{2} \text{ cosec } \alpha$$

Frictional resistance,

$$f = \mu N + \mu N = 2\mu N$$
$$= 2\ \mu \times \frac{R}{2} \text{ cosec } \alpha$$
$$= \mu R \text{ cosec } \alpha$$
$$= R \times \mu \text{ cosec } \alpha.$$
$$\therefore \quad \frac{T_1}{T_2} = e^{(\mu \text{ cosec } \alpha)\theta}$$
$$= e^{\mu\theta \text{ cosec } \alpha}$$

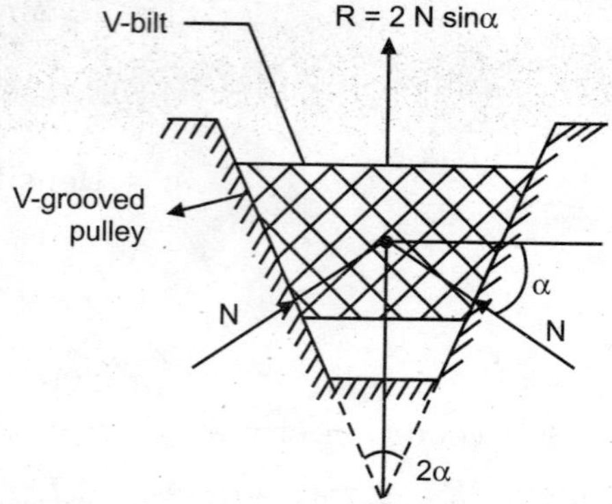

Fig. 6.13. V-belt

Example 6.9. **The maximum allowable tension in a V-belt of groove angle 45° is 1500 N. The belt is running at 2 m/s and angle of lap is 170°. Take μ = 0.27. Determine.**

(*a*) Net driving tension

(*b*) Power transmitted.

Solution :

$$2\alpha = 45°$$
$$\alpha = 22.5°$$
$$T_{max} = 1500 \text{ N}$$
$$\theta = 170° = 170 \times \frac{\pi}{180} = 2.967 \text{ rad.}$$
$$\mu = 0.27$$
$$v = 2 \text{ m/s}$$
$$T_{max} = T_1 + T_C = 1500 \qquad (\because T_C = 0)$$
$$\frac{T_1}{T_2} = e^{\mu\theta \text{ cosec } \alpha} = e^{0.27 \times 2.967 \text{ cosec } 22.5°}$$
$$= 8.1$$
$$T_2 = \frac{T_1}{8.1} = \frac{1500}{8.1} = 185 \text{ N.}$$

(*a*) Net driving tension $= T_1 - T_2$

$= 1500 - 185 =$ **1315 N Ans.**

(*b*) Power transmitted,

$$P = \frac{(T_1 - T_2) \times v}{1000} = \frac{(1500 - 185) \times 2}{1000}$$

= **2.63 kW Ans.**

QUESTION BANK

1. Explain briefly various belt and pulley drive arrangements. Give their applicaitons.
2. What is velocity ratio? Find velocity ratio for an open belt drive and cross belt drive.
3. What is stepped pulley drive? What conditions must be satisfied in its design?
4. Write notes on the following:
   (*a*) Fast and loose pulley drive
   (*b*) Belt drive with idler pulleys.
5. Find an expression for the length of belt for open belt drive. How is it different for corss belt drive?
6. Derive an expression for ratio of tensions for flat belt drive. How is it different for V-belt drive?
7. What are centrifugal tension and initial tension? How do these affect the maximum tension in a belt?
8. What are different conditions for maximum power transmission?

NUMERICAL PROBLEMS

1. A belt transmits power between two pulleys running at 31.67 m/s. The angle of lap is 165° and coefficient of function is 0.3. The mass of the belt is 0.9 kg/m length and maximum permissible stress in the belt is 2 MN/m^2. Find
   (*i*) Maximum power transmitted.
   (*ii*) Initial tension is a flat belt of 100 mm width and 8 mm thickness.
   [**Ans.** (*i*) 14.82 kW (*ii*) 1.322 kN]
2. A V-belt of mass 0.9 kg/m is used to transmit power between two shafts. The angle of lap is 170° and angle of groove is 45°. Take $\mu = 0.17$ and maximum permissible tension as 2.2 kN. Find.
   (*i*) Velocity of belt for maximum power
   (*ii*) Power transmitted at this velocity. [**Ans.** (*i*) 28.54 m/s (*ii*) 30.67 kW]
3. The maximum permissible stress in a belt is 1.4 N/mm^2 and ratio of tension is 2.0. Find the maximum power transmitted by a belt 150 mm × 10 mm if its density is 1 Mg/m^3. [**Ans.** 14.75 kW]
4. Two parallel shafts 12 m apart are to be connected by a belt running over pulleys of diameter 480 cm and 80 cm respectively. Determine the length of the belt for
   (*i*) open belt drive
   (*ii*) cross belt drive [**Ans.** (*i*) 33.13 m (*ii*) 33.45 m]
5. A shaft rotating at constant speed of 160 rpm is connected by belting to parallel shaft 72 cm apart which has to run at 60, 80 and 100 rpm. The smallest pulley on the drive shaft is 4 cm in radius. Calculate the radii of remaining stepped pulleys. [**Ans.** r_2 = 10.67 cm; r_3 = 4.89 cm, r_4 = 9.78 cm r_5 = 5.64 cm and r_6 = 9.02 cm]

❏❏❏

CHAPTER 7

BRAKES AND DYNAMOMETERS

7.1. INTRODUCTION

A brake is an arrangement to apply a retarding force in the form of a frictional resistance to a moving body to reduce or completely stop the motion. The brake absorbs the kinetic energy of the motion and converts the same into heat. The heat is dissipated to the atmosphere directly or through a cooling medium such as air, water, oil or other coolant. Brakes are important for the control and safe operation of the machines and plants.

There are many types of brakes used such as mechanical brakes, hydraulic brakes, electric brakes. The hydraulic and electric brakes are mostly used where large amounts of kinetic energy are to be transformed, i.e., locomotives, automobiles, dynamometer. The mechanical brakes of main types are:

1. Shoe brake or block brake
2. Simple band brake
3. Combined bank and block brake
4. Internal expanding shoe brake.

A dynamometer is a brake with additional facility to measure the frictional resistance applied. The torque transmitted or power developed by the machine at a given speed can be calculated from the frictional resistance measured by the dynamometer. There are mainly two types of dynamometer.

1. Absorption dynamometers absorb the power produced by the machine by applying friction resistance.
2. Transmission dynamometers donot absorb the power produced which is used for doing work by belt transmission or tosion of a shaft.

The difference between a brake and a dynamometer is that of purpose rather than of principle. A brake may be used as a dynamometer by addition of a force-measuring device such as a spring balance.

7.2. SHOE BRAKE

A shoe brake has a block of a soft material which is pressed against the rim of a rotating brake wheel or drum. The force on the drum is increased by using a lever. These brakes are used on railway trains and tram cars. Wood and rubber blocks are used for light and slow vehicles whereas cast iron or steel is used for heavy and fast vehicles. The material of block is always softer than that of drum or rim of the wheel to avoid wear of the drum.

7.2.1 Single Block Shoe Brake

A single block shoe brake with block rigidly fixed to the lever is shown in Fig. 7.1. Force P is applied at one end of the lever which is pivoted on a fixed fulcrum O. The angle of contact between block and wheel is assumed less than 60° and the normal pressure between block and wheel is taken uniform.

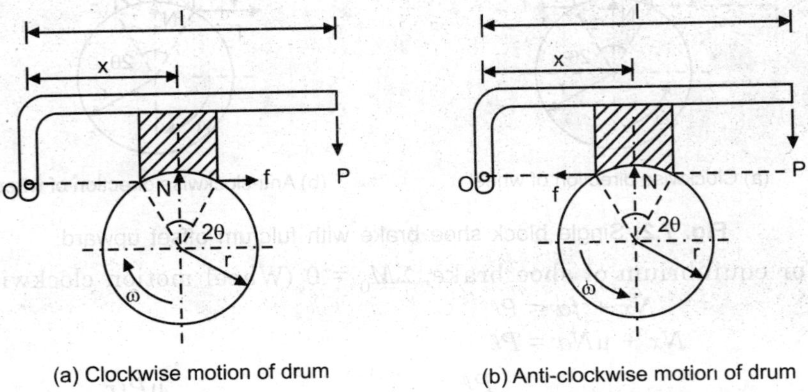

Fig. 7.1. Single Block Shoe Brake.

When P = Force applied at the end of the lever

N = Normal force between block and wheel

f = Force of friction or tangential braking force

μ = Coefficient of friction

2θ = Angle of contact of block surface

r = radius of wheel

$\therefore \quad f = \mu N$

The braking toque,

$$T = fr = \mu N r$$

There can be three cases for considering the equilibrium of the shoe brake.

Case 1. The tangential braking force (*f*) passes through fulcrum (*O*)

The system is shown in Fig. 7.1.

For equilibrium of the shoe brake, $\Sigma M_0 = 0$

$$\therefore \quad Nx = P\ell$$

or
$$N = \frac{P\ell}{x}$$

Braking torque, $T = \mu N r = \mu \times \dfrac{P\ell}{x} \times r = \dfrac{\mu P \ell r}{x}$

The braking troque is same independent of direction of rotation of the wheel.

Case 2. The line of action of tangential braking force (*f*) is at a distance '*a*' below the fulcrum, *O*.

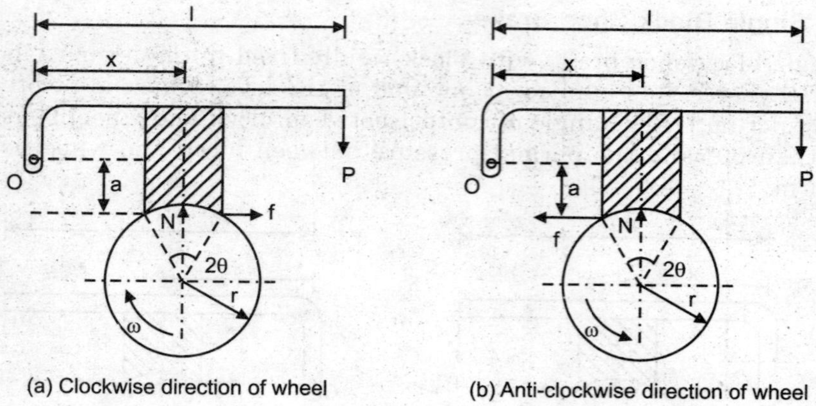

Fig. 7.2. Single block shoe brake with fulcrum offset upward.

For equilibrium of shoe brake, $\Sigma M_0 = 0$ (Wheel motion clockwise)

$$Nx + fa = P\ell$$

$$Nx + \mu Na = P\ell$$

$$\therefore \quad N = \frac{P\ell}{x + \mu a} \quad \text{or} \quad T = \mu r N = \frac{\mu P \ell r}{x + \mu a}$$

For equilibrium of shoe brake, $\Sigma M_0 = 0$ (Wheel motion anti-clockwise)

$$Nx - fa = P\ell$$

$$Nx - \mu Na = P\ell$$

$$\therefore \quad N = \frac{P\ell}{x - \mu a}$$

$$T = \mu N r = \frac{\mu P \ell r}{x - \mu a}$$

Case 3. The line of action of tangential braking force (*f*) is at a distance '*a*' above line of fulcrum, *O*.

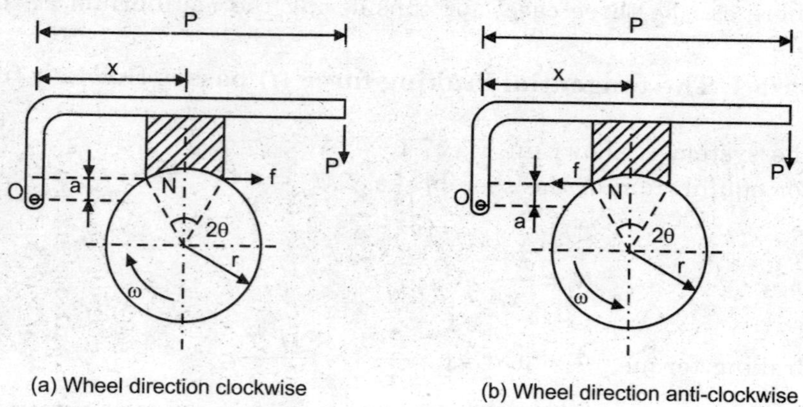

Fig. 7.3. Single block shoe brake with fulcrum offset downwards.

For equilibrium of shoe brake, $\Sigma M_0 = 0$ (Wheel motion clockwise)

$$Nx - fa = P\ell$$

$$Nx - \mu Na = P\ell$$

$$\therefore \quad N = \frac{P\ell}{x - \mu a}$$

$$T = \mu Nr = \frac{\mu P\ell r}{x - \mu a}$$

For equilibrium of shoe brake, $\Sigma M_0 = 0$ (Wheel motion anti-clockwise)

$$Nx + fa = P\ell$$

$$Nx + \mu Na = P\ell$$

$$\therefore \quad N = \frac{P\ell}{x + \mu a}$$

$$T = \mu Nr = \frac{\mu P\ell r}{x + \mu a}$$

Observations

1. Self-energizing brakes

When the brake wheel rotates anticlockwise in case 2 and clockwise in case-3,

$$Nx = P\ell + \mu Na \qquad \text{...(1)}$$

The frictional force (f) helps to apply brakes and such brakes are called self-energizing brakes. If f is sufficient to apply brakes with external force $P = 0$, the brakes are called **self-locking brakes**.

In equation (1), put $P = 0$

$$\therefore \quad Nx = \mu Na$$

or

$$x = \mu a$$

This is the condition for self-locking brakes. These brakes are used in back-stop application.

2. The brakes should be self-energizing and not self-locking. In order to avoid self-locking.

$$x > \mu a$$

3. The braking pressure on the shoe,

$$p = \frac{N}{A}$$

where A = projected area of the block $= (2r \sin\theta) \times b$

where b = width of the shoe

$2r \sin\theta$ = projected length of shoe

4. When a single block shoe brake is used, there is additional load (N) on shaft bearing which may cause bending of the shaft. In order to overcome the above drawback, double block shoe brake is used.

Example 7.1. **A single block shoe brake is shown where the block is rigidly fixed to the lever *AB*. The radius of drum is 160 mm and coefficient of friction of brake lining is 0.3. If the drum rotates clockwise, calculate the brake torque due to vertical force of 600 N applied at *B*.**

Solution : $r = 160$ mm $= 0.16$ m

$\mu = 0.3$

$P = 600$ N

$2\theta = 40°$, the block pressure may be taken as uniform

For equilibrium of brake shoe,

$$\Sigma M_A = 0$$

$$N \times 350 + f(200 - 160) = 600\,(400 + 350)$$

$$N = \frac{f}{\mu}$$

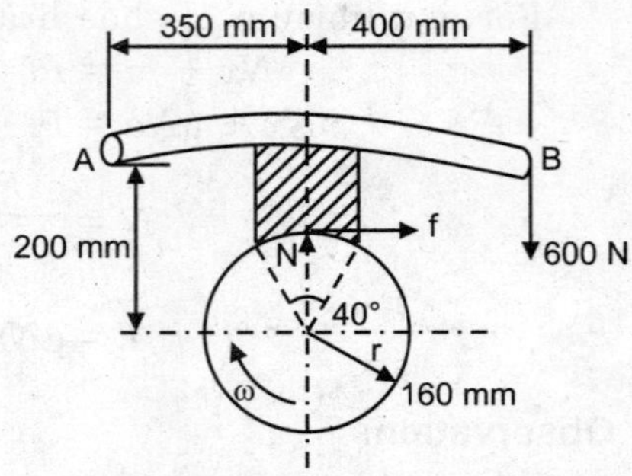

Fig. 7.4

$$\therefore \quad \frac{f}{0.3} \times 350 + 40\,f = 600 \times 750$$

$$\therefore \quad f = \frac{600 \times 750}{\frac{350}{0.3} + 40} = 372.8 \text{ N}$$

The braking torque,

$$T = f \times r = 372.8 \times 0.16 = \textbf{59.6 Nm Ans.}$$

7.2.2. Double Block Shoe Brake

A double block shoe brake is shown in Fig. 7.5 in which two brake blocks are applied at the opposite ends of the rim of the wheel to reduce the unbalanced force on the shaft. The braking action is doubled by the use of two blocks which can be operated by the same force applied by an electromagnet or a solenoid. This type of brake is used in cranes.

The braking torque,

$$T = (f_1 + f_2)r$$

where f_1 and f_2 are the braking force on the two blocks.

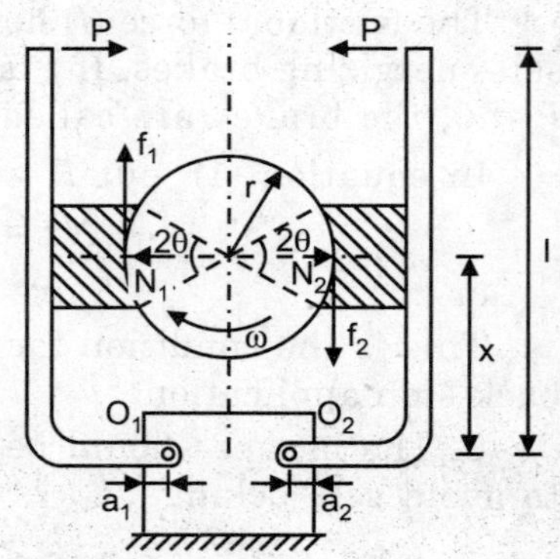

Fig. 7.5. Double Block Shoe Brake

7.2.3. Pivoted Block Shoe Brake

The normal pressure between the block and wheel is assumed uniform for short blocks with angle of contact less than 60°. For blocks with angle of contact move than 60°, the normal pressure between block and wheel is non-uniform with more pressure in the centre than at the ends of the block. In order to ensure uniform pressure and shoe wear, the block is hinged or pivoted to the lever instead of attaching rigidly.

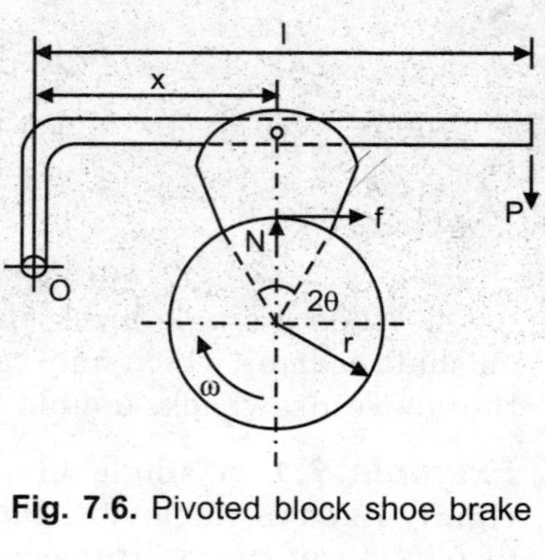

Fig. 7.6. Pivoted block shoe brake

The braking torque,

$$T = f \times r = \mu' N r$$

where μ = actual coefficient of friction

μ' = equivalent coefficent of friction

$$= \frac{4\mu \sin\theta}{2\theta + \sin 2\theta}$$

Pivoted block shoe brakes wear uniformly, have more life and higher braking torque than shoe brakes with rigid blocks.

Example 7.2. **A single block shoe brake has drum diameter of 250 mm and coefficient of friction of 0.35. The angle of contact is 90° and operating force is 700 N at the end of the lever shown. Determine the torque which can be transmitted by the brake.**

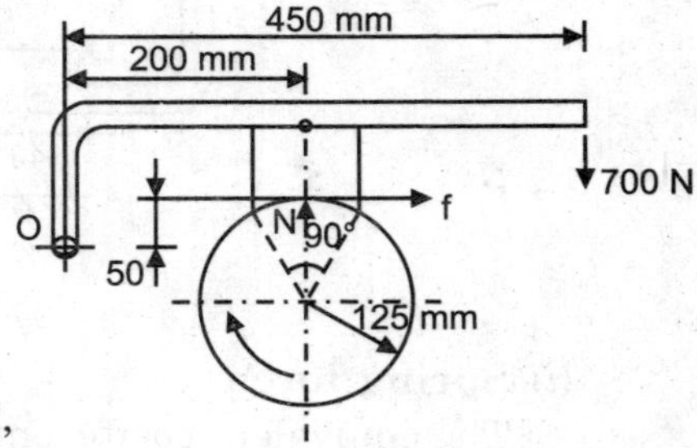

Fig. 7.7

Solution : $d = 250$ mm

$r = 125 \text{ mm} = 0.125 \text{ m}$

$$2\theta = 90° = \frac{\pi}{2}$$

$P = 700$ N

$\mu = 0.35$

The angle of contact is more than 60°, therefore equivalent coefficient of friction is used

$$\mu' = \frac{4\mu\sin\theta}{2\theta+\sin 2\theta} = \frac{4\times 0.35 \sin 45°}{\frac{\pi}{2} + \sin 90°} = 0.385.$$

For equilibrium of shoe brake $\Sigma M_0 = 0$

$$700 \times 450 + f \times 50 = N \times 200$$

$$= \frac{f}{\mu'} \times 200 = \frac{f}{0.385} \times 200 = 520\, f$$

$\therefore$ $520\, f - 50\, f = 700 \times 450$

$\therefore$ $$f = \frac{700 \times 450}{470} = 670 \text{ N}$$

The torque transmitted,

$$T = f \times r = 670 \times 0.125 = \mathbf{83.75 \text{ Nm}} \textbf{ Ans.}$$

Example 7.3. **A double shoe brake can absorb a torque of 1400 Nm. The brake has a drum diameter of 350 mm, angle of contact of each shoe is 100° and coefficient of friction is 0.4. The various dimension are shown in Fig. 7.8. Calculate.**

(*i*) the spring force necessary to set the braking, and

(*ii*) the width of the brake shoes if the bearing pressure on the lining material is not to exceed 0.3 N/mm².

Solution : $T = 1400 \text{ Nm} = 1400 \times 10^3$ N–mm

$d = 350$ mm

$\therefore$ $r = 175$ mm

$$2\theta = 100° = 100 \times \frac{\pi}{180} = 1.75 \text{ rad}$$

$\mu = 0.4$

$p = 0.3 \text{ N/mm}^2$

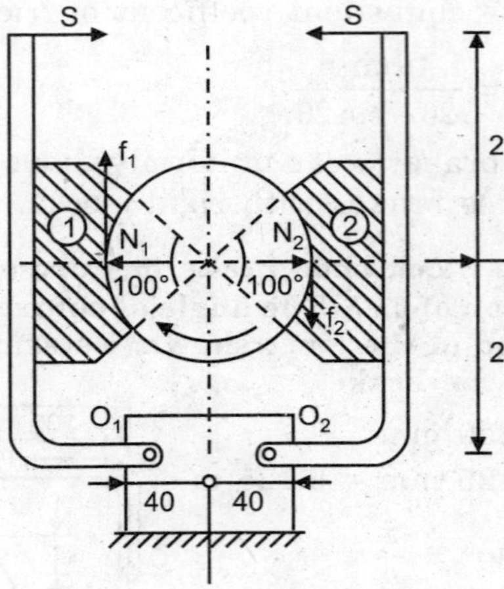

Fig. 7.8

(*i*) **Spring force**

The equivalent coefficient of friction (as angle of contact is more than 60°),

$$\mu' = \frac{4\mu \sin\theta}{2\theta + \sin 2\theta} = \frac{4 \times 0.4 \sin 50°}{1.75 + \sin 100°}$$

$$= 0.45$$

For equilibrium of shoe 1, $\Sigma M_{O_1} = 0$

$$S \times 450 + f_1\,(175 - 40) = N_1 \times 200$$

$$= \frac{f_1}{0.45} \times 200 = 444.4\, f_1$$

$$\therefore \quad 444.4\, f_1 - 135\, f_1 = S \times 450$$

$$\therefore \quad f_1 = \frac{S \times 450}{(444.4 - 135)} = 1.454\, S$$

For equilibrium of shoe 2, $\Sigma M_{o_2} = 0_2$.

$$S \times 450 = N_2 \times 200 + f_2(175 - 40)$$

$$= \frac{f_2}{0.45} \times 200 + f_2 \times 135 = 579.4\, f_2.$$

$$\therefore \quad f_2 = \frac{S \times 450}{579.4}$$

$$= 0.776\, S$$

Total torque capacity

$$T = 1400 \times 10^3 = (f_1 + f_2)r$$

$$= (0.776\ \text{s} + 1.454\ \text{s}) \times 175 = 390.25\, S$$

$$\therefore \quad S = \frac{1400 \times 10^3}{390.25}$$

$$= \mathbf{3587\ N} \quad \textbf{Ans.}$$

(*ii*) **Width of shoe brake**

Projected area for one shoe,

$$A = (2\ r \sin \theta) \times b$$

where b = width of shoe brake

Normal force on shoe 1,

$$N_1 = \frac{f_1}{\mu'} = \frac{1.454\ S}{0.45} = \frac{1.454 \times 3587}{0.45} = 11590\ \text{N}$$

Normal force on shoe 2,

$$N_2 = \frac{f_2}{\mu'} = \frac{0.776 S}{0.45} = \frac{0.776 \times 3587}{0.45} = 6186\ \text{N}$$

The maximum force is on shoe 1,

$$\therefore \quad p = \frac{N_1}{A}$$

$$0.3 = \frac{11590}{(2 \times 175 \sin 50°) \times b} = \frac{11590}{268b} = \frac{43.25}{b}$$

$$\therefore \quad b = \frac{43.25}{0.3} = \mathbf{144.2\ mm} \quad \textbf{Ans.}$$

7.3. BAND BRAKE

A band brake consists of a flexible band of leather or steel lined with friction material or a number of ropes which embrace a part of drum circumference. The band is tightened on the drum of the brake with the help of a simple lever or a differential lever. The band brakes are accordingly called simple band brake and differential band brake.

7.3.1. Simple Band Brake

One end of band is attached to a fixed pin (fulcrum) of the lever and the other end is attached at some distance from the fulcrum. The arrangement is shown in Fig. 7.9.

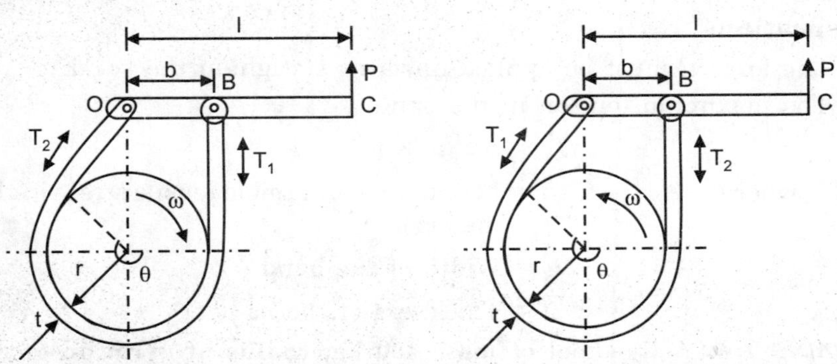

(a) Clockwise rotation of drum (b) Anti-clockwise rotation of drum

Fig. 7.9. Simple Band Brake

When force P is applied, the lever turns about fulcrum O and band is tightened and brakes are applied. The friction between band and drum provides the braking force.

when, T_1 = Tension in the tight side of band

T_2 = Tension is the slack side of band

θ = Angle of lap (or embrace) of band and drum

μ = Coefficient of friction between band and drum

r = Radius of brake drum

t = Thickness of band

r_e = Effective radius of drum $= r+\frac{t}{2}$

The ratio of tensions in the band,

$$\frac{T_1}{T_2} = e^{\mu\theta}$$

or,

$$\ln\left(\frac{T_1}{T_2}\right) = \mu\theta$$

Braking force on drum $= (T_1 - T_2)$

Braking torque on drum $= (T_1 - T_2)r_e$

$= (T_1 - T_2)r$ (Neglecting thickness of band)

Case I. Clockwise Rotation of Drum

The end of band attached to fulcrum is slack with tension T_2 and other end of band attached to point B will be tight with tension T_1.

For equilibrium of brake, $\Sigma M_0 = 0$

$$P \times l = T_1 \times b$$

Case II. Anticlockwise Rotation of Drum

The end of band attached to fulcrum will be tight with tension T_1 and other end attached to point B will be slack with tension T_2.

For equilibrium of brake, $\Sigma M_0 = 0$

$$\therefore \quad P \times l = T_2 \times b$$

Observations

1. The force P must be applied upwards to tighten the band.
2. The maximum tension in the band,

$$T_1 = \sigma \times w \times t$$

where σ = Maximum permissible tensile stress in the band material

w = Width of the band

t = Thickness of the band.

Example 7.4. **A flywheel of mass 400 kg, radius of gyration of 450 mm runs at 300 rpm. A simple band brake as shown is applied to the shaft of flywheel. The brake drum has a diameter of 240 mm and coefficient of friction is 0.2. Find:**

(*i*) the torque applied due to hand load of 100 N,

(*ii*) the number of turns of shaft before it comes to rest, and

(*iii*) the time required to bring the shaft to rest after application of brake.

Solution :

$m = 400$ kg

$k = 450$ mm

$= 0.45$ m

$N = 300$ rpm

$$\omega = \frac{2\pi N}{60} = \frac{2\pi \times 300}{60}$$

$= 31.42$ rad/s.

$\mu = 0.2$

$d = 240$ mm

$r = 120$ mm $= 0.12$ m

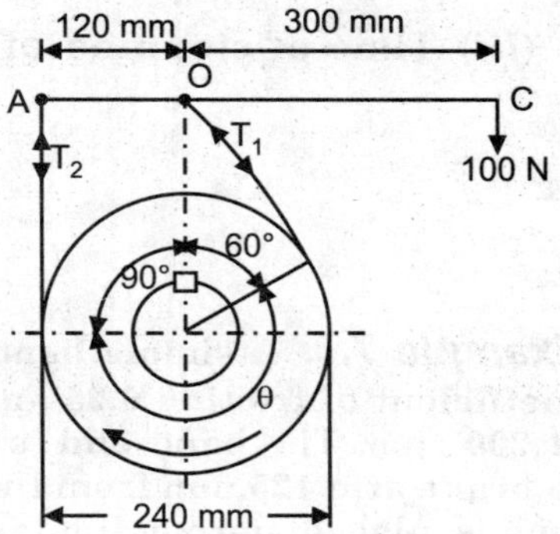

Fig. 7.10.

(*i*) Torque applied

From the direction of brake drum as clockwise, the tensions in the band as T_1 and T_2 are shown in the Fig. 7.10.

The angle of embrace,

$$\theta = 360° - (90° + 60°) = 210° = 210 \times \frac{\pi}{180}$$

$$= 3.666 \text{ rad}$$

$$\frac{T_1}{T_2} = e^{\mu\theta}$$

$$\frac{T_1}{T_2} = e^{0.2 \times 3.666} = 2.08$$

For equilibrium of brake, $\Sigma M_0 = 0$

$$T_2 \times 120 = 100 \times 300$$

$$T_2 = \frac{100 \times 300}{120} = 250 \text{ N}$$

$$T_1 = 2.08;\ T_2 = 2.08 \times 250 = 520 \text{ N.}$$

The braking torque applied,

$$T = (T_1 - T_2)\, r = (520 - 250) \times 0.12$$

$= \mathbf{32.4}$ **Nm Ans.**

(*ii*) Number of turns of shaft before coming to rest (N)

The kinetic energy of rotation of flywheel,

$$\text{KE} = \frac{1}{2} \times I\omega^2 = \frac{1}{2} \times mk^2\omega^2$$

$$= \frac{1}{2} \times 400 \times (0.45)^2 (31.42)^2 = 40{,}000 \text{ Nm}$$

The work done by braking torque,

$$\text{W.D.} = 2\pi NT = 2\pi N \times 32.4 = 203.6 \text{ N}$$

$$\text{KE} = \text{WD}$$

$\therefore$ $203.6 \text{ N} = 40{,}000$

$\therefore$ $N = \dfrac{40{,}000}{203.6} = \mathbf{196.5}$ **Ans.**

***(iii)* Time of stopping of shaft**

$$N = \frac{n - n_0}{t} = \frac{300 - 8}{t}$$

$$\therefore \qquad t = \frac{N}{300} = \frac{196.5}{300} = \mathbf{0.655\ min\ Ans.}$$

Example 7.5. **A simple band brake has a drum of 600 mm diameter and coefficient of friction 0.25 and angle of contact 270°. The drum is rotating at 200 rpm. The band end is connected to the fulcrum pin and other end to brake arm 125 mm from the pin. The straight brake arm is 750 mm long and is placed perpendicular to the diameter that bisects the angle of contact. Find:**

(*i*) the necessary pull to be applied at the end of brake arm to step the wheel if 35 kW is being absorbed. What is the direction of the pull?

(*ii*) The width of the steel band if the band thickness is 2.5 mm and maximum allowable tensile stress is not to exceed 250 N/mm².

Solution :

$d = 600$ mm

$r = 300$ mm $= 0.3$ m

$N = 200$ rpm

$\mu = 0.25$

$\theta = 270°$

$= \frac{270 \times \pi}{180}$

$= 4.713$ rad

Power $= 35$ kW

$= 35 \times 10^3$ W

$t = 2.5$ mm

$\sigma = 50$ N/mm²

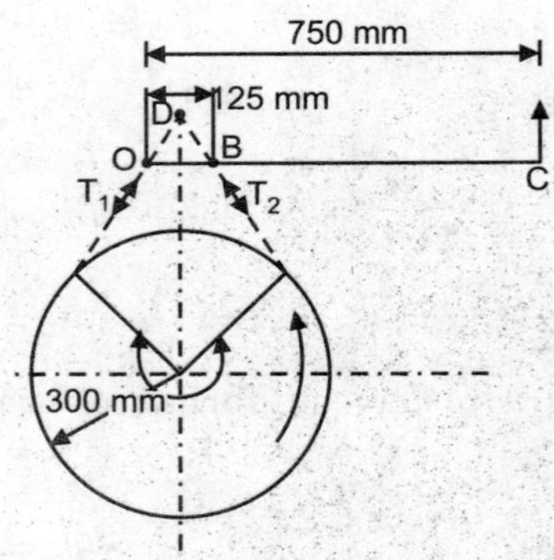

Fig. 7.11

(*i*) Pull (*P*) required

For anticlockwise direction of motion, the band tensions T_1 and T_2 are shown in Fig. 7.11. The pull has to be applied upwards

$$\frac{T_1}{T_2} = e^{\mu\theta} = e^{0.25 \times 4.713}$$

$$\therefore \qquad \frac{T_1}{T_2} = 3.25$$

$$\text{The power absorbed} = \frac{2\pi NT}{60}$$

$$\text{The braking torque, } T = \frac{35 \times 10^3 \times 60}{2\pi \times 200} = 1667 \times 10^3 \text{ Nmm}$$

$$\text{The braking torque, } T = (T_1 - T_2)\, r$$

$$\therefore \qquad (T_1 - T_2) = \frac{1667 \times 10^3}{300} = 5556 \text{ N.}$$

$$\left(T_1 - \frac{T_1}{3.25}\right) = 5556 \text{ N}$$

$$\therefore \quad T_1 = 8025 \text{ N}$$

$$T_2 = \frac{8025}{3.25} = 2469 \text{ N}$$

For equilibrium of brake, $\Sigma M_0 = 0$

$$\therefore \quad P \times 750 = T_2 \times OD$$

where
$$OD = \frac{125}{2} \times \frac{1}{\cos 45°} = 62.5 \times \sqrt{2}$$

$$\therefore \quad P = \frac{2469 \times 62.5 \times \sqrt{2}}{750} = \mathbf{291 \text{ N Ans.}}$$

(ii) Width of steel band

$$T_1 = \sigma w t$$

The width of steel band

$$w = \frac{T_1}{\sigma t} = \frac{8025}{50 \times 2.5} = \mathbf{64.2 \text{ mm Ans.}}$$

7.3.2. Differential Band Brake

A special brake arm is used where both ends of the band are fixed to the arm at different distances from the fulcrum and no end is fixed to the fulcrum. The distance of slack end from fulcrum is more than distance of tight end to ensure proper lightening of the band.

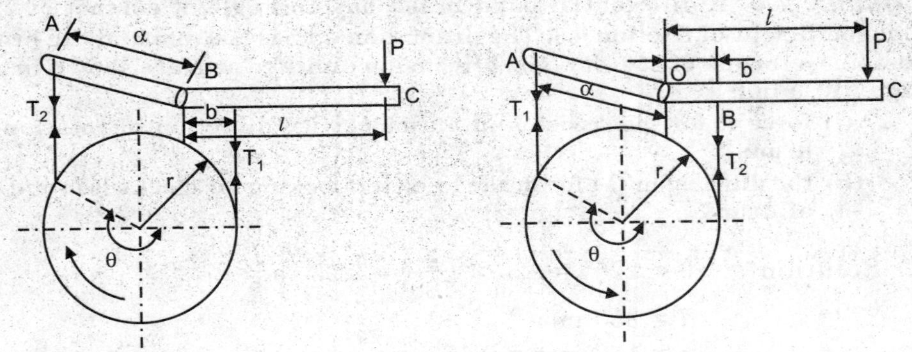

(a) Clockwise rotation of drum (b) Anti-clockwise rotation of drum

Fig. 7.12. Differential Band Brake

The braking torque is obtained in similar way as in a simple band brake.

***(a)* Clockwise rotation of drum**

The band end attached to end A is slack with tension T_2 and end attached to B is tight with tension T_1.

$\Sigma M_0 = 0 \quad P \times l + T_1 b = T_2 a \quad$ or $Pl = T_2 a - T_1 b$

***(b)* Anticlockwise rotation of drum**

The band end attached to end A is tight with tension T_1 and end attached to B is slack with tension T_2

$\Sigma M_0 = 0 \quad P \times l + T_2 b = T_1 a \quad$ or $Pl = T_1 a - T_2 b$

Observations

1. For self energizing brake, the moments T_1b and T_2b help Pl to apply brake for clockwise and anticlockwise rotation of drum respectively.
2. Condition for self-locking brake.

(*i*) Clockwise motion

$$T_2a \le T_1b$$

or $$\frac{T_2}{T_1} \le \frac{b}{a}$$

or $$\frac{T_1}{T_2} \le \frac{b}{a}$$

(*ii*) Anticlockwise motion

$$T_1a \le T_2b$$

or $$\frac{T_1}{T_2} \le \frac{b}{a}$$

or $$\frac{T_2}{T_1} \ge \frac{a}{b}$$

3. The length OA is greater than OB, the force P is applied downward. If length OB is greater than OA, the force P is applied upward.

The tensions T_1 and T_2 in the band remain unchanged.

Example 7.6. **A differential band brake has an angle of contact of 225° and coefficient of friction 0.3. The drum diameter is 350 mm and the brake sustains a torque of 350 Nm. The dimension of brake arm are shown in the Fig. 7.13. Find:**

(*i*) force *P* for clockwise and anticlockwise direction of rotation of drum.

(*ii*) the dimension *OA* for brake to be self-locking for clockwise rotation of drum.

Solution : $\theta = 225° = 225 \times \frac{\pi}{180} = 3.93 \text{ rad}$

$d = 350$ mm

$r = 175$ mm $= 0.175$ m

$\mu = 0.3$

$T = 350$ Nm $= 35 \times 10^3$ Nmm

$$\frac{T_1}{T_2} = e^{\mu\theta} = e^{0.3 \times 393} = 3.255$$

Braking torque

$$T = (T_1 - T_2)r$$

$$(T_1 - T_2) = \frac{T}{r} = \frac{350 \times 10^3}{175} = 2000 \text{ N.}$$

$$T_1 = 2887 \text{ N}$$

$$T_2 = 887 \text{ N}$$

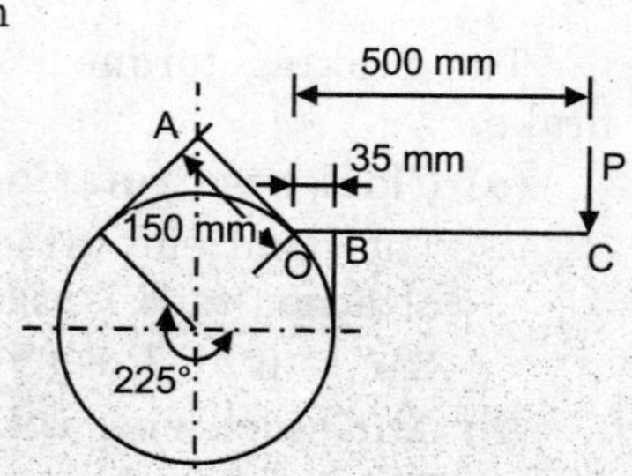

Fig. 7.13

(i) Calculation of applied force P

(a) Clockwise motion of drum

$$\Sigma M_0 = 0$$

$$P \times 500 = T_2 \times 150 - T_1 \times 35$$

$$\therefore \quad P = \frac{887 \times 150 - 2887 \times 35}{500}$$

$$= \mathbf{64\ N\ Ans.}$$

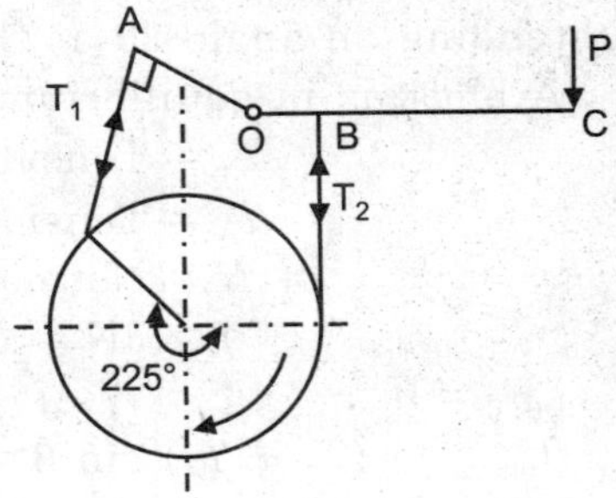

Fig. 7.14

(b) Anticlockwise motion of drum

$$\Sigma M_0 = 0$$

$$P \times 500 = T_1 \times 150 - T_2 \times 35$$

$$\therefore \quad P = \frac{2887 \times 150 - 887 \times 35}{500}$$

$$= \mathbf{804\ N\ Ans.}$$

(ii) Value of OA for self-locking brake

For clockwise rotation of drum,

$$P \times 500 = T_2 \times OA - T_1 \times OB$$

For self-locking brake,

$$P = 0$$

$$\therefore T_2 \times OA = T_1 \times OB$$

$$\therefore \quad OA = \frac{T_1}{T_2} \times OB = 3.255 \times 35$$

$$= \mathbf{114\ mm\ Ans.}$$

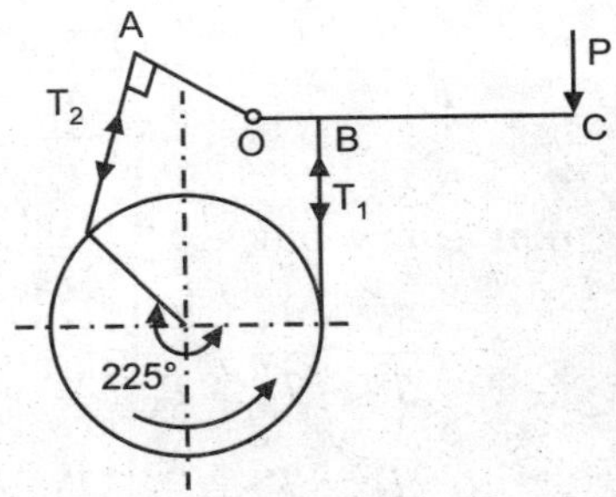

Fig. 7.15

7.4. BAND AND BLOCK BRAKE

The blocks are made of wood or other friction material to ensure high coefficient of friction and facility to replace them easily on wear. A number of such blocks are attached inside a steel band to make a band and block brake. The braking force can be more continently applied to a band.

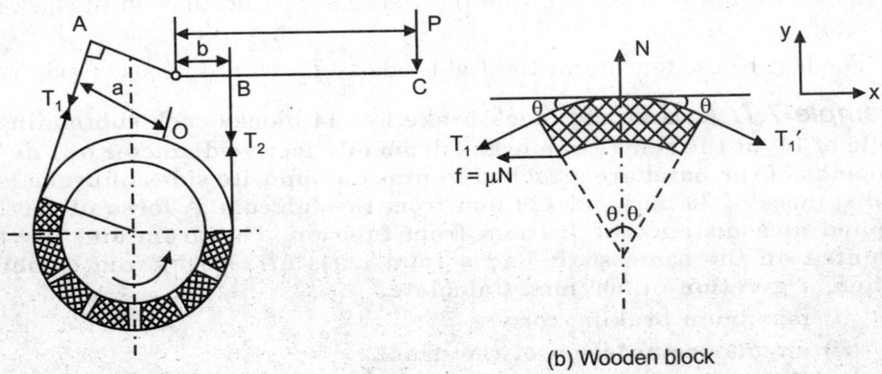

(a) Brake

(b) Wooden block

Fig. 7.16. Band and Block Brake

A band and block brake shown in Fig. 7.16 has n wooden block each subtending an angle 2θ at the drum centre.

A block is in equilibrium under the following forces.

T_1 = Tension on tight side

T_1' = Tension on slack side

N = Normal reaction of the drum on the block

f = μN = force of friction

$\Sigma F_y = 0;$

$$(T_1 + T_1') \sin \theta = N$$

$\Sigma F_x = 0;$

$$(T_1 - T_1') \cos \theta = \mu N$$

$$\therefore \quad \frac{(T_1 - T_1')\cos\theta}{(T_1 + T_1')\sin\theta} = \frac{\mu N}{N}$$

$$\therefore \quad (T_1 - T_1') = \mu \tan \theta \, (T_1 + T_1')$$

$$\therefore \quad \frac{T_1}{T_1'} = \frac{1+\mu\tan\theta}{1-\mu\tan\theta}$$

For every block,

$$\frac{T_1'}{T_2'} = \frac{T_2'}{T_3'} = \frac{T_3'}{T_4'} = \text{.........} \frac{T_{n-1}}{T_2} = \frac{1+\mu\tan\theta}{1-\mu\tan\theta}$$

$$\frac{T_1}{T_2} = \frac{T_1}{T_1'} \times \frac{T_1'}{T_2'} \times \frac{T_2'}{T_3'} \times \text{......} \times \frac{T_{n-1}}{T_2} = \left(\frac{1+\mu\tan\theta}{1-\mu\tan\theta}\right)^n$$

The braking torque on the drum,

$$T = (T_1 - T_2)\, r_e$$

$$= (T_1 - T_2)\, r \quad \text{(Neglecting thickness of band)}$$

Observations

1. For block 1, the tension in the tight side is T_1 and tension on slack side is T_1'.
2. For block 2, the tension in the tight side is T_1' and tension in slack side is T_2'
3. For last block, tension in the tight side is T_{n-1} and in slack side is T_2.

Example 7.7. **A band and block brake has 14 blocks each subtending an angle of 15° at the centre of a brake drum of effective diameter of 1 m. The two ends of the band are attached to pins on opposite sides of brake lever at distances of 30 mm and 120 mm from the fulcrum. A force of 200 N is applied at a distance of 750 mm from fulcrum. The drum and flywheel mounted on the same shaft had a total mass of 2000 kg and combined radius of gyration of 500 mm. Calculate.**

(*i*) maximum braking torque

(*ii*) angular retardation of the drum

(*iii*) time taken by the system to come to rest from a rated speed of 360 rpm.

The coefficient of friction between blocks and drum way be taken as 0.25.

Solution:

$$n = 14$$
$$2\theta = 15°$$
or
$$\theta = 7.5°$$
$$d = 1 \text{ m}$$
$$r = 0.5 \text{ m} = 500 \text{ mm}$$
$$m = 2000 \text{ kg}$$
$$k = 500 \text{ mm} = 0.5 \text{ m}$$
$$P = 200 \text{ N}$$
$$l = 750 \text{ mm}$$
$$\mu = 0.25$$

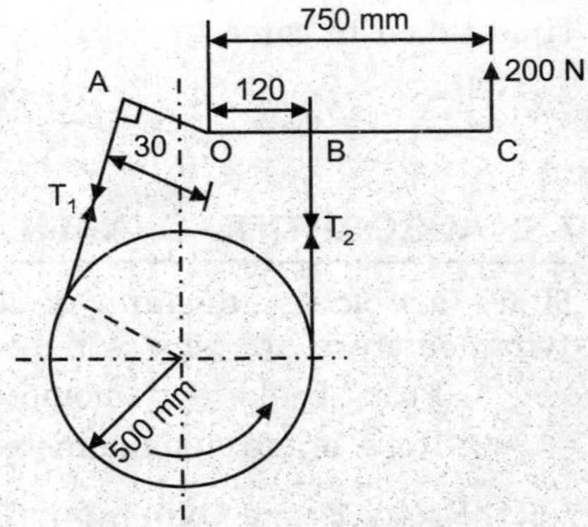

Fig. 7.17

(i) Maximum Braking Torque

For maximum braking torque, the direction of rotation should be anticlockwise and $OB > OA$. The force P must he applied upward and band end attached to A has tight tension T_1 and end attached to B has slack tension T_2.

For eqilibrium of brake arm, $\Sigma M_0 = 0$

$$P \times l + T_1 \times a = T_2 \times b$$
$$12 \times 750 + T_1 \times 30 = T_2 \times 120$$
$$12T_2 - 3T_1 = 15000 \quad ...(i)$$

Also
$$\frac{T_1}{T_2} = \left(\frac{1+\mu\tan\theta}{1-\mu\tan\theta}\right)^n$$
$$= \left(\frac{1+0.25\tan 7.5°}{1-0.25\tan 7.5°}\right)^{14} = 2.512 \quad ...(ii)$$

From equations (*i*) and (*ii*)

$$T_1 = 8440 \text{ N}$$
$$T_2 = 3360 \text{ N}$$

Therefore, maximum braking torque,

$$\tau = (T_1 - T_2)r = (8440 - 3360)\, 0.5 = \mathbf{2540\ Nm\ Ans.}$$

(ii) Angular retardation of drum

The torque $\tau = I\alpha = mk^2\alpha$

where angular retardation

$$\alpha = \frac{\tau}{mk^2} = \frac{2540}{2000(0.5)^2} = \mathbf{5.08\ rad/s^2\ Ans.}$$

(iii) Time taken to come to rest

Initial angular velocity

$$\omega_1 = \frac{2\pi N_1}{60} = \frac{2\pi \times 360}{60} = 37.7 \text{ rad/s}$$

Final angular velocity

$$\omega_2 = 0$$
$$\omega_2 = \omega_1 - \alpha t$$

Time taken to come to rest,

$$t = \frac{\omega_2 - \omega_1}{\alpha} = \frac{0 - 37.7}{5.08} = 7.42 \text{ s} \quad \textbf{Ans.}$$

7.5. ABSORPTION DYNAMOMETERS

There are many designs of absorption dynamometers. The following dynamometers are very widely used.

1. Proxy brake dynamometer
2. Rope brake dynamometer

7.5.1. Proxy Brake Dynamometer

A simple proxy brake dynamometer consists of two wooden blocks clamped together on a revolving pulley mounted on the engine shaft. It carries a lever with weights hanging on the other end, to counter balance the brake. The friction between the blocks when clamped tightly and pulley tends to rotate the blocks in the direction of rotation of the shaft. The weight W tries to counter balance the above tendency.

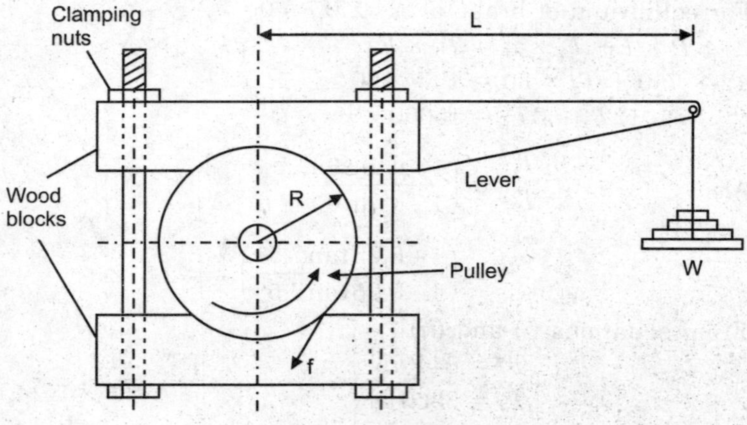

Fig. 7.18. Proxy Brake Dynamometer

The grip of the blocks is adjusted with the help of clamping nuts to allow the engine run at the required speed. The weights are added to keep the lever horizontal in equilibrium position, the power of the engine is absorbed by the friction.

Assume, W = Weight at the lever end (N)

L = Horizontal distance of weight from centre of the pulley [m]

f = Frictional resistance between blocks and pulley [N]

R = Radius of the pulley [m]

N = Shaft speed (rpm).

The frictional torque,

$$T = WL = = mgL \text{ [Nm]}$$

Brake power of the engine

$$P = \frac{2\pi NT}{60 \times 1000} [\text{kW}]$$

$$= \frac{2\pi NmgL}{60 \times 1000} = \text{kmN}$$

where $$k = \frac{2\pi gL}{60 \times 1000} = \text{constant for a given brake.}$$

Observations

1. The engine power is independent of size of pulley and coefficient of friction and amount of tightening of clamping nuts.
2. The dynamometer is subjected to severe oscillations when engine torque is not uniform.

7.5.2. Rope Brake Dynamometer

Rope brake dynamometer is widely used in measuring the brake power of the engine in the test laboratories. A rope is wrapped over the rim of a pulley mounted on the engine shaft. The rope is secured to the pulley with the help of 4 wooden blocks. The upper end of rope is attached to a spring balance and the lower end supports weight of suspended mass.

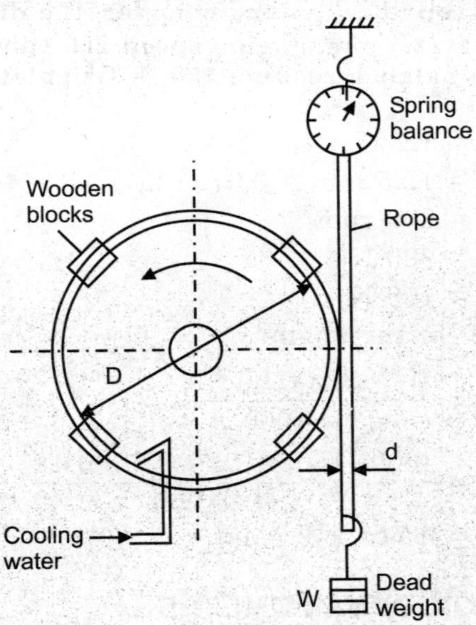

Fig. 7.19. Rope Brake Dynamometer

The engine is made to run at a constant speed, i.e., the frictional torque of the rope is equal to torque transmitted by the engine.

Assume, W = Dead weight [N]

S = Spring balance reading [N]

D = Wheel diameter [m]

d = Rope diameter [m]

N = Engine speed [rpm]

Net load on the engine = $(W - S)$ [N]

Work done per revolution = Net load × Distance moved

$$= (W - S) \times \pi(D + d) \quad \text{[Nm]}$$

Brake power of the engine

$$P = \frac{(W-S)\pi(D+d)N}{60\times1000} \quad \text{[kW]}$$

$$= k(W - S)\,N$$

where $k = \dfrac{\pi(D+d)}{60\times1000}$ = Constant for the dynamometer

Observations

1. The frictional resistance of the rope brake is transformed into heat which is removed with the help of water spray on the pulley.
2. The spring balance can be removed by wrapping a number of turns of the rope around the pulley.

$$\frac{T_1}{T_2} = e^{\mu\theta}$$

where $T_1 = W$

$T_2 = S$

Example 7.8. **In a rope brake dynamometer, the diameter of pulley was 1.2 m, rope diameter 12.5 mm, engine speed 200 rpm, dead weight on the brake 600 N, spring balance reading 150 N. Calculate the brake power of the engine.**

Solution : $D = 1.2$ m

$d = 12.5$ mm $= 0.0125$ m

$N = 200$ rpm

$W = 600$ N

$S = 150$ N

The brake power of the engine,

$$P = \frac{(W-S)\,\pi\,(D+d)N}{60\times1000}$$

$$= \frac{(600-150)\,\pi\,(1.2+0.0125)\times200}{60\times1000}$$

= **5.715 kW Ans.**

7.6. TRANSMISSION DYNAMOMETERS

There are many transmission dynamometers. The most widely used dynamometers are as follows.

1. **Belt transmission dynamometer**

   There are Froude Transmission Dynamometer or Thoneycraft Transmission Dynamometer.

2. **Torsion dynamometer**

   This includes Bevis-Gibson Flash Light Torsion Dynamometer.

7.6.1. Belt Transmission Dynamometer

A belt transmits power from one pulley to another which is proportional to difference in tensions between the tight and slack sides of the belt. Froude or Thoneycroft belt transmission dynamometer can measure the difference in tension $(T_1 - T_2)$ directly when the belt is running.

A continuous belt runs over the driving pulley (*A*) rigidly secured to the engine shaft and driven pulley (*B*) through two intermediate pulleys (*C* and *D*). The intermediate pulleys are fixed to a lever having a fulcrum (*E*) at the midpoint of the two pulley centres. A suspended weight at one end of a lever counterbalances the difference in the tensions of the tight and slack sides

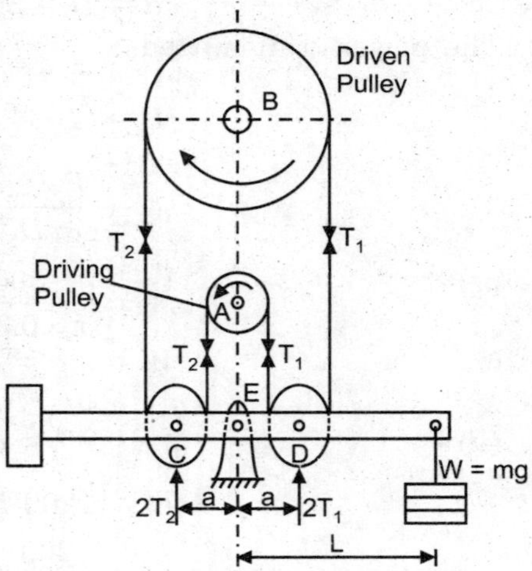

Fig. 7.20. Belt Transmission Dynamometer

Assume, T_1 = Tension on tight side of driving pulley [N]

T_2 = Tension on slack side of driving pulley [N]

$2T_1$ = Total force on pulley *D* [N]

$2T_2$ = Total force on pulley *C* [N]

W = Dead weight [N]

L = Distance between dead weight and fulcrum (*E*) [m]

For equilibrium of the lever, $\Sigma M_E = 0$ (neglecting friction)

$$2T_1 \times a = 2T_2 \times a + WL$$

or
$$(T_1 - T_2) = \frac{WL}{2a}$$

Assume D = Diameter of driving pulley, *A* [m]

N = Engine speed (rpm)

The brake power of the engine,

$$P = \frac{(T_1 - T_2)\pi DN}{60 \times 1000} \text{ [kW]}$$

$$= \frac{(T_1 - T_2)V}{60 \times 1000} \text{ kW}$$

where $V = \pi DN$ = Belt speed (m/s)

Example 7.9. **In a belt transmission dynamometer, the speed of driving pulley is 300 rpm. The diameter of both driving and driven pulleys is 360 mm. The distance of counter weight from centre of the pulleys is 800 mm.**

Find the value of counter weight required to maintain the lever in horizontal position where power transmitted is 3 kW. Also, find its value when the belt just begins to slip on the driving pulley. Take μ = 0.25 and maximum allowable tension in the belt as 1200 N.

Solution : (*i*)

$$P = 3 \text{ kW}$$
$$L = 800 \text{ mm} = 0.8 \text{ m}$$
$$N = 300 \text{ rpm}$$
$$a = D = 360 \text{ mm} = 0.36 \text{ m}$$

The power transmitted,

$$P = (T_1 - T_2)V = \frac{mgL}{2a} \times \frac{\pi ND}{60}$$

$$\therefore \quad m = \frac{P \times 2a \times 60}{\pi DNgL}$$

$$= \frac{3000 \times 2 \times 0.36 \times 60}{\pi \times 0.36 \times 300 \times 9.81 \times 0.8} = 48.7 \text{ kg.}$$

or $W = mg = 48.7 \times 9.81 = 477.75\ N.$

(*ii*)

$$\frac{T_1}{T_2} = e^{\mu\theta} = e^{0.25 \times \pi} = 2.19$$

$$T_1 = 1200 \text{ N}$$

$$T_2 = \frac{1200}{2.19} = 548 \text{ N}$$

$$(T_1 - T_2) = \frac{WL}{2a}$$

$$\therefore \quad W = \frac{(T_1 - T_2) \times 2a}{L} = \frac{(1200 - 548) \times 2 \times 0.36}{0.8}$$

$$= \mathbf{586.64\ N.\ Ans.}$$

7.6.2. Torsion Dynamometer

When power is transmitted by a shaft, the driving end of the shaft twists through a small angle relative to driven end of the shaft. The amount of twist can be measured from the torsion equation

$$\frac{T}{J} = \frac{G\theta}{L}$$

where T = Torque acting on the shaft [Nm]

J = Polar moment of inertia [m^4]

$$= \frac{\pi}{32} D^4 \quad \text{(for a solid shaft)}$$

$$= \frac{\pi}{32}(D^4 - d^4) \quad \text{[for a hollow shaft]}$$

θ = Angle of twist (radian)

G = Modulus of rigidity [N/m^2]

L = Length of the shaft [m]

The power transmitted,

$$P = \frac{2\pi NT}{60 \times 1000} \text{(kW)}$$

where N = speed of the shaft (rpm)

$$T = \frac{GJ}{L} \times \theta = k.\theta$$

where $k = \dfrac{GJ}{L}$ = constant for a given shaft.

Therefore, torque acting on the shaft is proportional to angle of twist. If angle of twist can be measured, the power transmitted by the shaft can be calculated.

In a Bevis-Gibson Torsion Dynamometer, two similar discs are mounted at a known distance on the shaft. The two discs revolve with the shaft. A lamp is masked and fixed on the shaft bearing. A moveable torque finder is used which has an eye piece capable of moving circumferentially. Each disc has a small radial slot near its periphery. Similar slots are also made at the same radius on the mask of the lamp and torque finder.

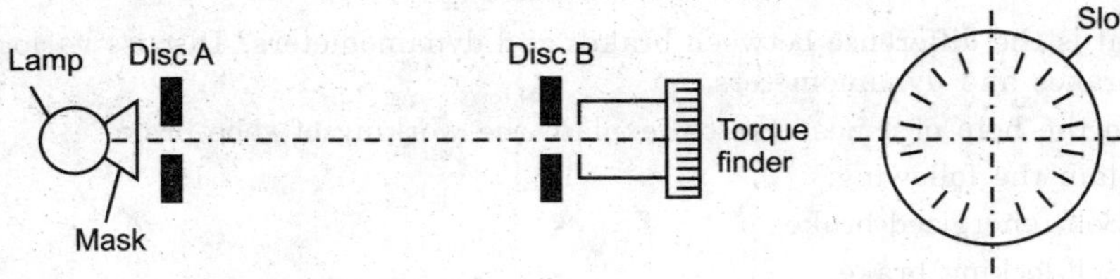

Fig. 7.21. Bevis-Gibson Torsion Dynamometer

1. When the shaft rotates freely and no torque is transmitted, all the four slots are in a line and a ray of light from the lamp can be seen through the eye piece of torque finder after every revolution.
2. When the torque is transmitted, the shaft twists and slot in disc B shifts in position and no light is visible through eye piece of torque finder.
3. The eye piece is moved circumferentially by the amount of shaft twist till light is visible again once in each revolution of shaft. The eye piece is moved by a micrometer spindle which measures the angle of the twist upto one hundredth of a degree.

Example 7.10. **A torsion dynamometer fitted to the propeller shaft of a marine engine shows shaft twist of 2° in a length of 20 m at a steed of 120 rpm. The shaft has external diameter of 400 mm and internal diameter of 300 mm. Find the engine power if G = 80 GPa.**

Solution : $\theta = 2° = 2 \times \dfrac{\pi}{180} = 0.035$ rad

$L = 20$ m

$N = 120$ rpm

$D = 400$ mm $= 0.4$ m

$d = 300$ mm $= 0.3$ m

$G = 80$ GPa $= 80 \times 10^9$ N/m^2

$$J = \frac{\pi}{32}(D^4 - d^4) = \frac{\pi}{32}\left[(0.4)^4 - (0.3)^4\right]$$
$$= 0.0017 \text{ m}^4$$

$$\frac{T}{J} = \frac{G\theta}{L}$$

$$\therefore \quad T = \frac{GJ}{L}\theta = \frac{80 \times 10^9 \times 0.0017 \times 0.035}{20}$$

$$= 238 \times 10^3 \text{ Nm}$$

The engine power, $$P = \frac{2\pi NT}{60 \times 1000} = \frac{2\pi \times 120 \times 238 \times 10^3}{60 \times 1000}$$

$$= 2990 \text{ kW Ans.}$$

QUESTION BANK

1. What is the difference between brakes and dynamometers? Discuss various type of brakes and dynamometers.
2. With the help of a neat sketch, explain the working of shoe brake.
3. Explain the following:
   (*a*) Self- energized brake
   (*b*) Self-locking brake.
4. Discuss the effectiveness of band brake under various operating conditions.
5. Explain the working of a band and block brake. What is the relation for ratio of tight and slack side tensions?
6. Describe the construction and operation of a prony brake or rope brake absorption dynamometer.
7. Explain with detail the calculations involved in finding the power transmitted in a torsion dynamometer.
8. Explain the operation of Bevis-Gibson flash light dynamometer.

OBJECTIVE TYPE QUESTIONS

Tick (✓) the right answer:

1. Brakes commonly used in trains are
   (*a*) band (*b*) shoe
   (*c*) band and block (*d*) internal
2. In self-locking brake, the force required to apply the brake is
   (*a*) minimum (*b*) zero
   (*c*) maximum
3. When the frictional force helps the applied force, the brake is called:
   (*a*) self-locking (*b*) automatic
   (*c*) self-energizing

Key [1. (*b*); 2. (*b*); 3. (*c*)]

NUMERICAL PROBLEMS

1. A single block brake is running at 400 rpm in clockwise direction. The drum diameter is 800 mm and brake arrangement is shown in Fig. 7.2(*a*). The force required at the end of lever is 450 N. If angle of contact is 25° and $L = 1.5$ m, $a = 450$ mm and $b = 77.5$ mm, determine the braking torque. Take $\mu = 0.3$.
[**Ans.** 175.6 Nm]

2. The figure shows a simple band brake which is applied to shaft carrying a flywheel of mass 300 kg and radius of gyration 280 mm. The drum diameter is 220 mm and the shaft speed is 240 rpm. Coefficient of friction is 0.3. Find the brake torque when a force 100 N is applied at the lever end. Also determine the number of turns of the flywheel and time taken by it before rest. All dimensions are in mm **[UPTU 2010-11]**

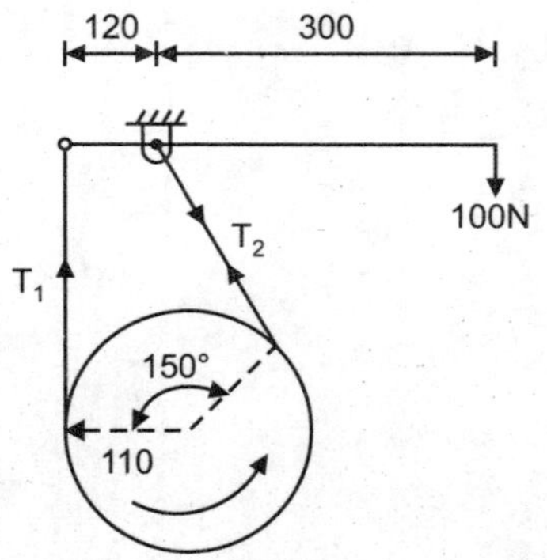

Fig. 7.22.

3. A torsion dynamometer is fitted on a turbine shaft to measure the angle of twist. It is observed that the shaft twists 1.5° in a length of 5 m at 500 rpm. The shaft is solid and has a diameter of 200 mm. If modulus of rigidly of shaft material is 85 GPa, find the power transmitted by the turbine. [**Ans.** 3662 kW]

❑❑❑

UNIT–IV

8. CAMS

Chapter 8

CAMS

8.1 INTRODUCTION

A cam is a mechanical element which drives another element called follower imparting a specific motion through direct contact. The cam may be rotating or reciprocating whereas the motion of the follower may be reciprocating, oscillating or of almost any desired characteristics. A cam and follower have contact and therefore, belong to the category of higher pairs. The cams are manufactured by die-casting, milling or by punch presses.

Cams are widely used in industry . These are used in automatic machines, internal combustion engines, machine tools, printing, control mechanisms, paper cutting mechanisms, spinning and weaving textile machineries, feed mechanism of automatic lathes, etc. The cam and follower is one of the most important mechanism found in modern machinery today due to the following special features.

1. Cam-and-follower mechanisms are simple and inexpensive.
2. They have very few moving parts.
3. The mechanism occupies a very small space.
4. It is not difficult to design the mechanism for any type of motion.
5. The most attractive feature of this mechanism is its versatility and flexibility.

However, a wide variety of shapes and forms are needed to meet various requirements of different applications. Therefore, there is a need of special terminology to distinguish different types of cams and followers.

8.2 DISPLACEMENT DIAGRAM FOR A CAM

There are a wide variety of cams used but there are certain common features. A cam system is a single-degree-of-freedom device. It is driver by a known input motion in the form of a shaft rotating at constant speed. A special shape is given to the cam to produce a certain desired periodic output motion of the follower.

A cam-follower mechanism has three elements, i.e., cam, follower and a guide or frame which acts as support of the assembly. During the rotation of the cam through one cycle (360°) of input motion, the follower executes a series of events as shown in displacement diagram of Fig. 8.1. The input motion θ is drawn to a scale on x-axis and the follower travel y is drawn to full scale on y-axis. The graph can be divided into the following parts.

1. **Rise:** The motion of the follower is away from the cam center. The maximum rise is called the **lift**. There is positive contact between, the follower and the cam.
2. **Dwell:** The portion of the cycle during which the follower is at rest is called the dwell.
3. **Return:** The portion of the cycle during which the motion of the follower is towards the cam center is called return.

During dwell and return portion of the cycle, the motion of the follower is not positively controlled by the cam and external force is provided by a spring.

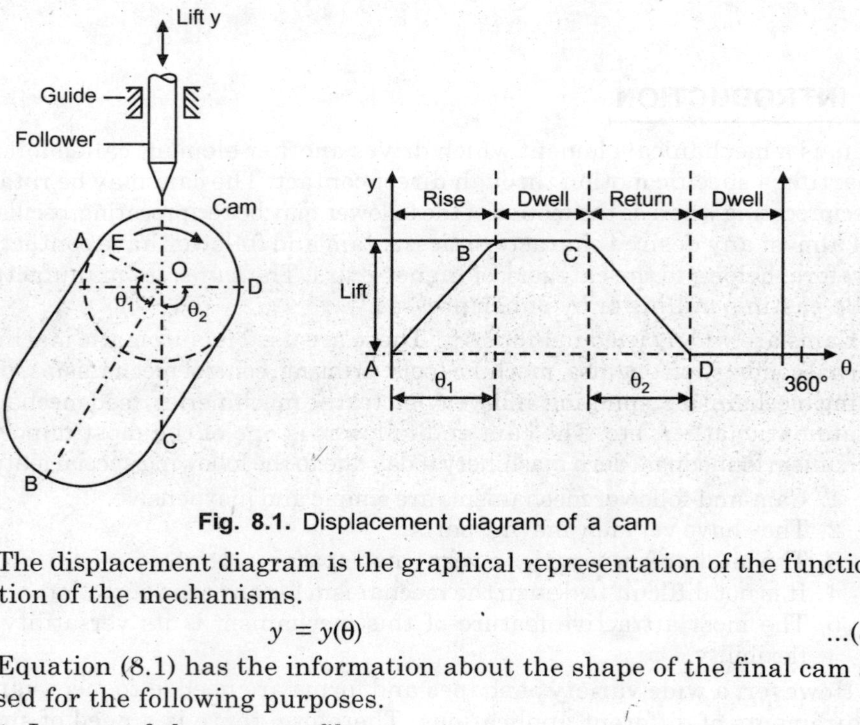

Fig. 8.1. Displacement diagram of a cam

The displacement diagram is the graphical representation of the functional relation of the mechanisms.

$$y = y(\theta) \qquad \text{...(8.1)}$$

Equation (8.1) has the information about the shape of the final cam and is used for the following purposes.

1. Layout of the cam profile.
2. Cam manufacture
3. Determination of the quality of dynamic preformance.

8.3. CLASSIFICATION OF CAMS AND FOLLOWERS

Cam system can be classified on the basis of different features.

8.3.1 Cam Shapes

Cams are classified according to their basic shapes. Four different types of cams are shown in Fig. 8.2.

Fig. 8.2(*a*) Plate cam, disc cam or radial cam
Fig. 8.2(*b*) A wedge cam
Fig. 8.2(*c*) Cylindrical cam or barrel cam
Fig. 8.2(*d*) End cam or face cam

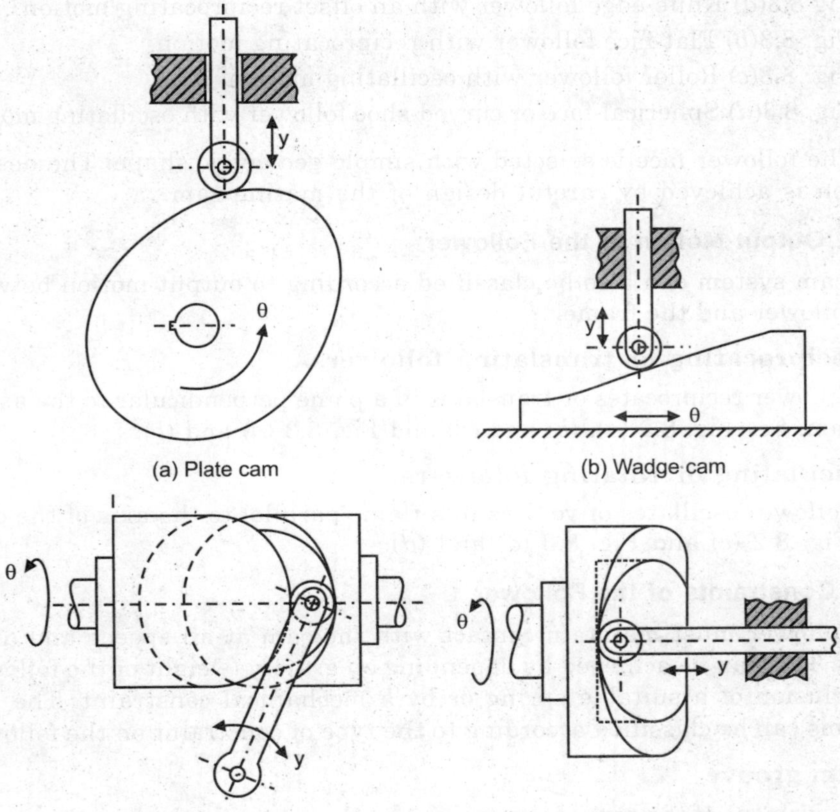

(a) Plate cam (b) Wadge cam (c) Cylindrical (d) Face cam

Fig. 8.2 Types of cam shapes

The wedge cam needs a reciprocating motion and not a continuous input motion. Therefore, this type of cam has very limited practical applications:

Plate cam is the most common type of cam.

8.3.2 Follower Shapes

Cam systems can also be classified according to basic shape of the follower. Four different system of follower are shown in Fig. 8.3 acting on plate cams.

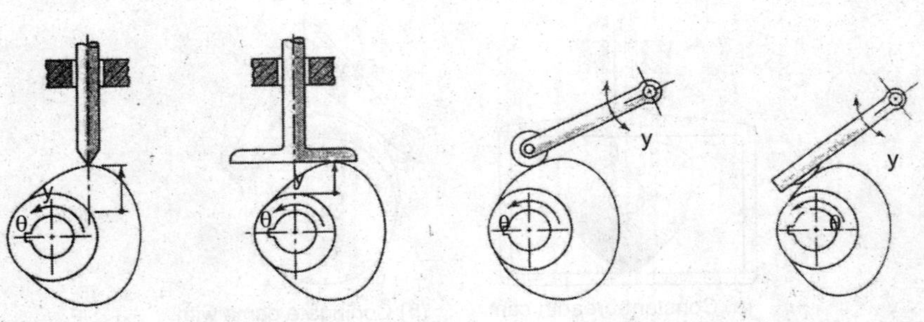

(a) Knife edge (b) Flat face (c) Roller (d) Curved-shoe

Fig. 8.3 Types of followers shapes

Fig 8.3(*a*) Knife-edge follower with an offset reciprocating motion.

Fig. 8.3(*b*) Flat-face follower with reciprocating motion.

Fig. 8.3(*c*) Roller follower with oscillating motion.

Fig. 8.3(*d*) Spherical-face or curved-shoe follower with oscillating motion.

The follower face is selected with simple geometric shape. The desired motion is achieved by careful design of the mating cam.

8.3.3. Output Motion of the Follower

The cam system can also be classified according to output motion between the follower and the frame.

1. Reciprocating or translating followers.

The follower reciprocates or translates in a plane perpendicular to the axis of the cam. See Fig. 8.2 (*a*), (*b*) and (*d*) and Fig. 8.3 (*a*) and (*b*).

2. Oscillating or rotating followers.

The follower oscillates or rotates in a flame parallel to the axis of the cam. See Fig. 8.2 (*c*) and Fig. 8.3 (*a*) and (*d*).

8.3.4 Constraints of the Follower

The follower must maintain contact with the cam at all speeds and at all times. This can be achieved by depending on gravity (weight of the follower) or inclusion of a suitable spring or by a mechanical constraint. The cam systems can be classified according to the type of constraint on the follower.

1. Cam groove

The movement of the cam is constrained by the groove in the barrel cam. See Fig. 8.2 (*c*)

2. Constant-breadth cam

Two contact points (at top and bottom) between the cam and the follower provide the constraint. See Fig. 8.4 (*a*).

3. Dual or conjugate cam

The arrangement has two cams with a common follower. However, each cam has its own roller. See Fig. 8.4 (*b*) which shows a conjugate cam with an oscillating roller follower.

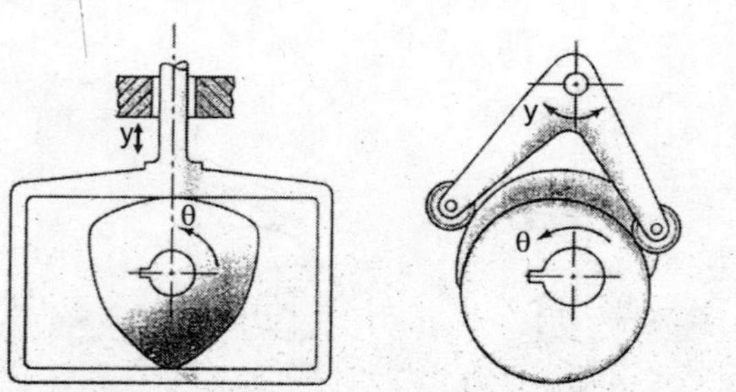

(a) Constant-breadth cam with a reciprocating flat-faced follower

(b) Conjugate cams with oscillating roller follower

Fig. 8.4. Contraints on the followe

5. Follower Displacement Motion

The displacement, velocity and acceleration diagrams of the cam system will depend upon the type of motion of the follower. The following types of motions are popular is the design of cam systems.

1. Uniform motion or uniform velocity
2. Uniform acceleration motion
3. Simple harmonic motion
4. Cycloidal motion.
5. Parabolic motion.

8.4. CAM NOMENCLATURE

8.4.1 Cam Profile

The following terminology is used in drawing the profile of a cam. The radial cam with reciprocating roller follower is most widely used and the same cam system is taken for defining the cam profile.

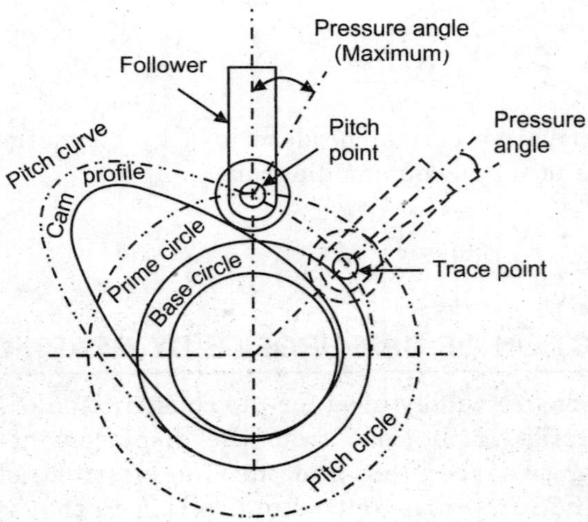

Fig. 8.5 Cam Nomenclature

1. **Base circle** is the smallest circle tangent to the cam (profile) drawn from the center of rotation of a cam shaft.
2. **Prime circle** is the smallest circle drawn tangent to curve.
3. **Pitch circle** is the circle passing through the pitch point and concentric with the base circle.
4. **Pitch curve** is the curve drawn by the trace point assuming that the cam is fixed and trace point of the follower rotates around the cam.
5. **Trace point** is a reference point on the follower to trace the cam profile with the center of the roller of a roller follower.

6. **Pitch point** is the point of the pitch curve at which the pressure angle is maximum.
7. **Presence angle** represents the steepness of the cam profile. It is the angle between the normal to the pitch curve at a point and the direction of the follower motion. The pressure angle varies in magnitude at all instances of the follower motion. The value of pressure angle should be small as large value can jam the follower in the bearings.

8.4.2. Kinematic Derivatives

The functional relation of cam system can be written as follows. (Equation 8.1)

$$y = y(\theta)$$

where y = displacement of the follower

θ = cam angle

The first derivative is obtained by differentiating with respect to θ.

$$\therefore \quad \dot{y}(\theta) = \frac{dy}{d\theta} = \text{slope of the displacement curve}$$

The second derivative gives the radius of curvature of the cam at different points.

$$\ddot{y}(\theta) = \frac{d^2 y}{d\theta^2}$$

The third derivative cannot be described by geometry but affects the relation of shape of displacement diagram.

$$\dddot{y}(\theta) = \frac{d^3 y}{d\theta^3}.$$

8.5. CONSTRUCTION OF CAM PROFILE BY GRAPHICAL METHODS

Graphical methods are widely used for the construction of cam profile. The shape of cam profile is obtained from the displacement diagram of the follower. The cam is made fixed and the line of stroke of the follower is revolved round the cam in the opposite direction to that is which the cam actually turns on its axis. This is called **principle of kinematic inversion**. The profile of the common surfaces of contact of cam and follower are joined by a smooth curve to get the actual profile of the cam.

The graphical construction, therefore, consists of two steps.

1. Drawing of displacement diagram.
2. Drawing of cam profile.

8.5.1 Construction of Displacement Diagram

1. Select suitable horizontal scale and vertical scale.
2. Draw a horizontal line to represent 360°.
3. Divide horizontal line into rise, dwell, return, dwell as per angle of outstroke, angle of dwell after angle of return and angle of dwell after return stroke.

4. Draw a vertical line from the origin and cut a distance to represent the lift (stroke) of the follower in mm.
5. Complete the rectangle *AFGE* as shown in Fig. 8.6.

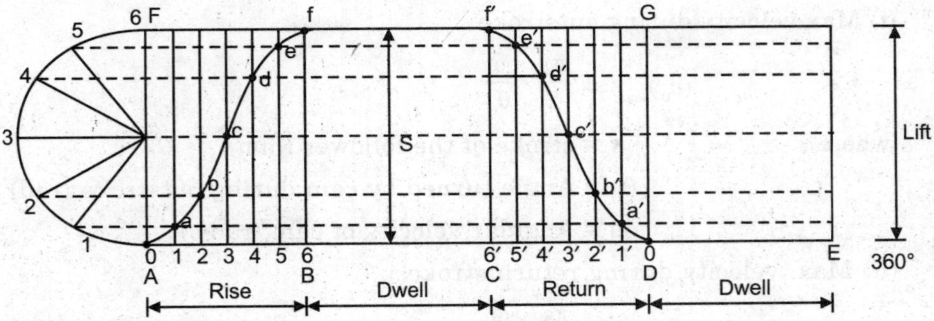

Fig. 8.6 Construction of Displacement Diagram

6. On diameter *AF* draw a semicircle and divide into parts (say 6) as 1, 2, 3, 4, 5, 6. Join these points with the centre of semi-circle.
7. Through every point of the semi-circle draw horizontal lines.
8. Divide the horizontal distance *AB* and *CD* into six equal parts. Draw vertical lines through these points intersecting the horizontal lines *a*, *b*, *c*, *d*, *e*, *f* and *a*′, *b*′, *c*′, *d*′, *e*′, *f*′.
9. Join these points by smooth curves *A-f-f′–D-E* is the required displacement diagram.

8.5.2 Construction of Cam Profile

The cam profile will depend open its direction of rotation of cam, the type and position of the follower. It is assumed that the cam rotates in clockwise direction and the axis of a knife edge follower passes through the axis of cam shaft.

1. Draw a prime circle with radius equal to minimum radius of the cam.
2. Mark points *A*, *B*, *C*, *D* and *A* to represent the rize, dwell, return and dwell angles.
3. Divide the angles *AOB* and *COD* into six parts each as was done for displacement diagram.
4. Draw radial lines 0-1, 0-2, 0-3, ---- 0-6 and 0-6', 0-5' 0-4' ---- 0-1'. Produce these radial lines.
5. Mark distances 1-*a*, 2-*b*, 3-*c*, ... 6-*f* and 6′-*f*′, 5′-*e*′, 4′-*d*′, ..., 1′-*a*′ corresponding to displacements by measurement from displacement diagram.
6. Draw a smooth curve *A*, *a*, *b*, *c*, *d*, *e*, *f* and *f*′, *e*′, *d*′, *c*′, *b*′, *a*′, *D*.
7. The smooth curve *A-f-f*′*-D -A* is the cam profile.

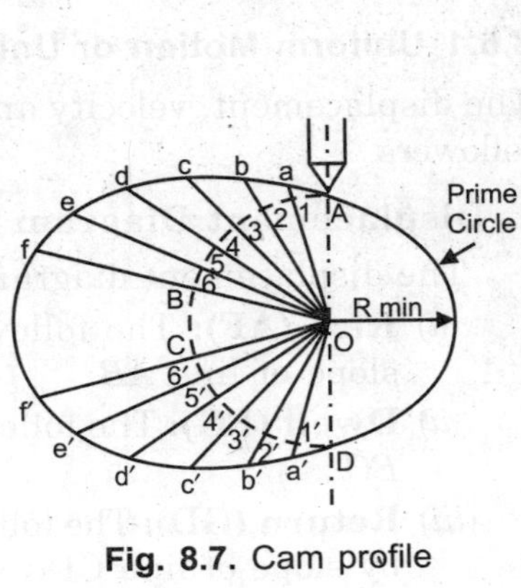

Fig. 8.7. Cam profile

8.5.3. Calculation of Maximum Velocity and Acceleration of Follower

These values will depend upon the type of motion. For simple harmonic motion;

(*i*) Max velocity during outstroke

$$(v_0)_{\max} = \frac{\pi s}{2} \times \frac{\omega}{\theta_0}$$

where s = stroke of the follower (mm)

θ_0 = Angle turned by cam during outstroke (rad)

ω = Angular velocity of cam (rad/s)

(*ii*) Max. velocity during return stroke.

$$(v_r)_{\max} = \frac{\pi s}{2} \times \frac{\omega}{\omega_r}$$

where ω_r = Angle turned by cam during return stroke (rad)

(*iii*) Max. acceleration during outstroke.

$$(a_0)_{\max} = \frac{\pi^2 s}{2} \times \frac{\omega^2}{\theta_0^2}$$

(*iv*) Max, acceleration during return stroke

$$(a_r)_{\max} = \frac{\pi^2 s}{2} \times \frac{\omega^2}{\omega_r^2}.$$

8.6 MOTIONS OF THE FOLLOWER

There may be many types of motions of the follower. The following motions are widely used.

1. Uniform motion or uniform velocity
2. Simple harmonic motion
3. Parabolic motion
4. Any type of desired motion by combination of above motions. The follower may be raised with uniform vélocity and lowered with simple harmonic motion.

8.6.1 Uniform Motion or Uniform Velocity of Follower

The displacement, velocity and acceleration diagrams can be constructed as followers.

1. Displacement Diagram

The displacement diagram can be divided into four periods.

(*i*) **Rise (AF):** The follower moves with uniform velocity upwards by slope of line *AB*.

(*ii*) **Dwell (FG):** The follower is at rest, represented by a horizontal line *BC*.

(*iii*) **Return (GD):** The follower returns with uniform velocity represented by slope of line CD.

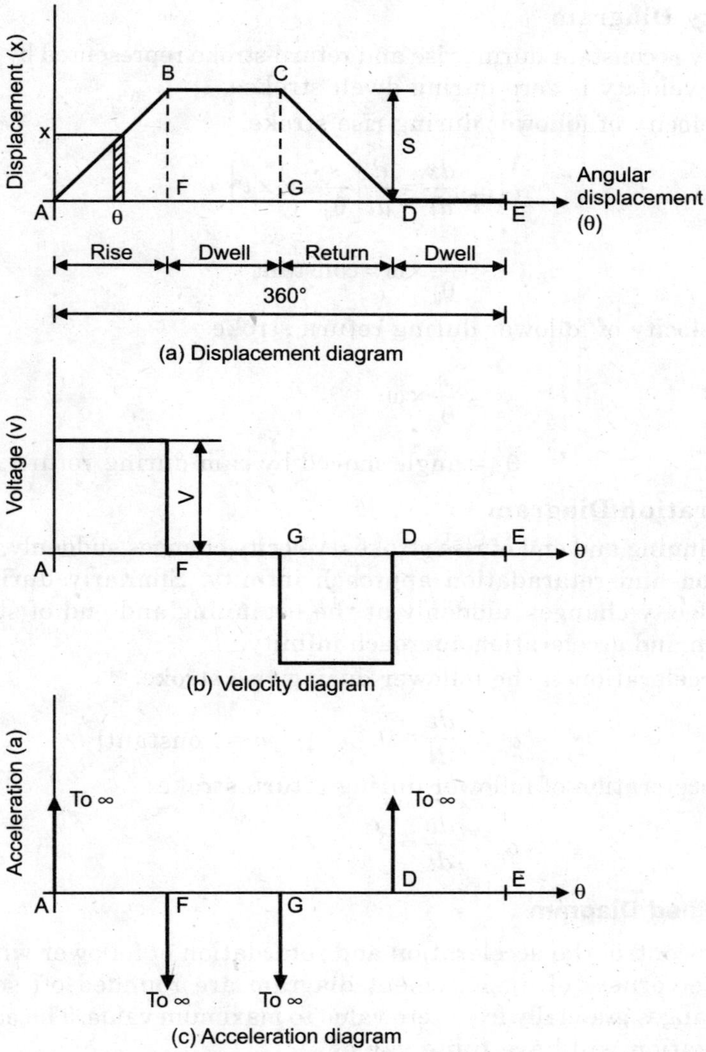

Fig. 8.8 Displacement, velocity, acceleration diagram for uniform motion

(*iv*) **Dwell (DF):** The follower is at rest represented by a horizontal line *DE*.

The angular displacement,

$$\theta = \omega \times t$$

where ω = angular velocity of cam

The displacement (lift) of follower

$$x = \frac{s}{\theta_0} \times \theta$$

where s = Max lift of the follower.

θ_0 = angle moved by cam during rise.

$$x = \frac{s}{\theta_0} \times \omega t$$

2. Velocity Diagram

The velocity is constant during rise and return stroke represented by horizontal lines. The velocity is zero during dwell strokes.

The velocity of follower during rise stroke.

$$v_o = \frac{dx}{dt} = \frac{d}{dt}\left[\frac{s}{\theta_0} \times \omega \times t\right]$$

$$= \frac{s}{\theta_0} \times \omega = \text{constant}$$

The velocity of follower during return stroke

$$v_r = \frac{s}{\theta_r} \times \omega$$

where θ_r = angle moved by cam during return stroke.

3. Acceleration Diagram

At the beginning and end of rise strokes, velocity changes suddenly, Therefore acceleration and retaradation approach infinity. Similarly during return stroke, velocity changes suddenly at the beginning and end of stroke. The retardation and acceleration approach infinity.

The acceleration of the follower during rise stroke.

$$a_o = \frac{dv}{dt} = 0 \qquad [\because v = \text{constant}]$$

The acceleration of follower during return stroke

$$a_r = \frac{dv}{dt} = 0$$

8.6.2 Modified Diagram

In order to control the acceleration and retardation of follower within finite values, the corners of displacement diagram are rounded off so that the velocity changes gradually from zero value to maximum value. The acceleration and retardation will have finite values.

The corners of the displacement diagram are rounded off by parabolic arcs.

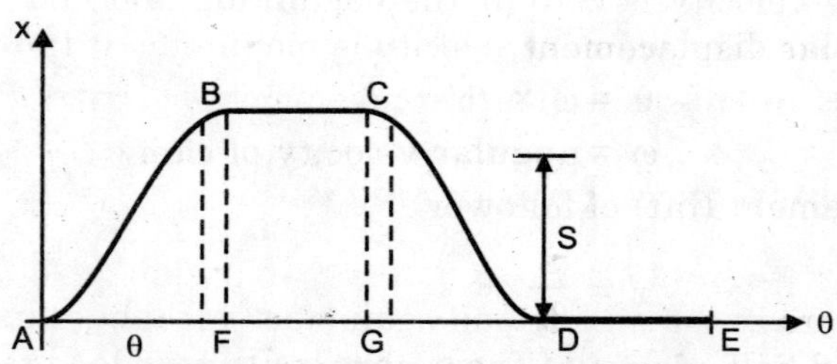

(a) Displacement diagram

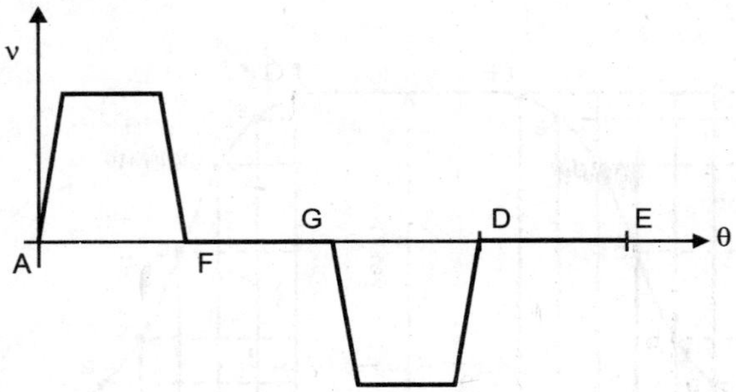

(b) Velocity diagram

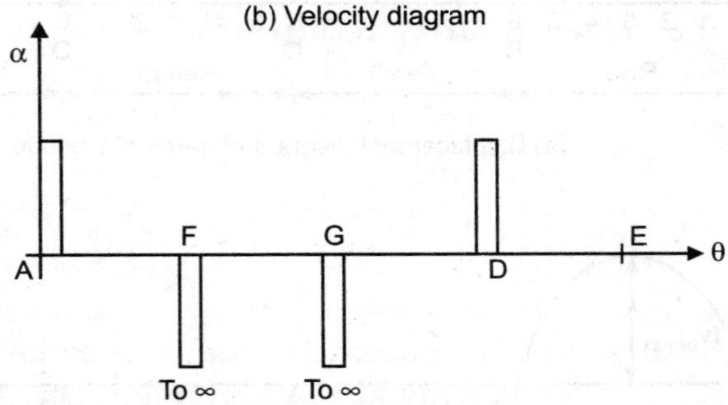

(c) Acceleration diagram

Fig. 8.9. Modified diagram for uniform motion

8.6.3 Simple Harmonic Motion

The displacement, velocity and acceleration diagrams for simple harmonic motion of follower can be drawn as follower.

1. **Displacement Diagram**
   - (*i*) Draw a semi-circle on stroke as diameter.
   - (*ii*) Divide the semi-circle into even number of equal parts.
   - (*iii*) Divide the angular displacement of cam during rise and return stroke of follower into same number of parts.
   - (*iv*) The points of intersection of horizontal projection from semi-circle points and vertical projection of angular displacements are joined by smooth curves to get complete displacement diagram. *A-F-G-D-E.*
2. **Velocity Diagram:** For harmonic motion, the velocity curve is a sine curve. The velocity is zero at the beginining and end of rise stroke and return stroke. The value of velocity is maximum at the middle of strokes.
3. **Acceleration Diagram:** For harmonic motion, the acceleration curve is a cosine curve. Acceleration is maximum at the beginning and at the end of rise and return strokes and zero in the middle of strokes.

> ***Note:*** There are abrupt changes of acceleration from maximum at beginning of stroke to zero in the middle and again abrupt changes in the deceleration during second half of stroke. These abrupt changes lead to jerks, noise and vibration, etc. The simple harmonic motion programme may be adopted for slow to medium speed. It is not suitable for high speed cams.

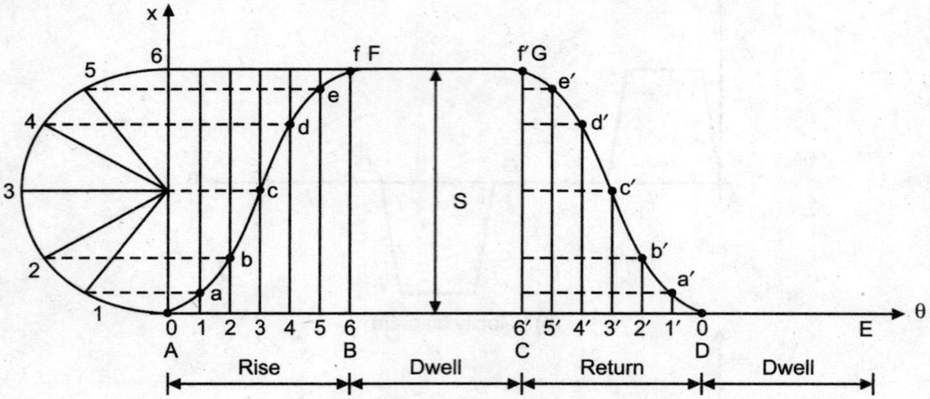

(a) Displacement diagram of harmonic motion

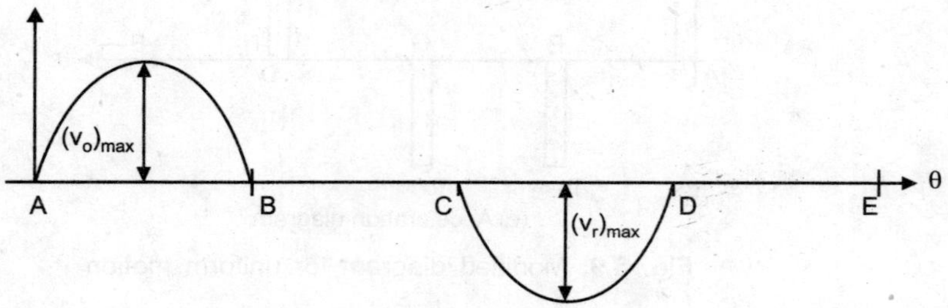

(b) Velocity diagram

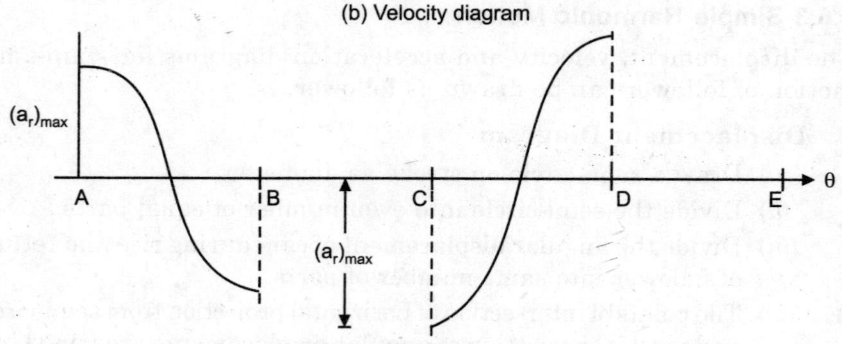

(c) Acceleration diagram

Fig. 8.10. Harmonic Motion

8.6.4. Analysis of Simple Harmonic Motion

Assume a point $R(\theta, x)$ moves along the semi-circle. When $\theta = \theta_0$, the angle turned by point R is π (180°)

$$\theta = \omega \times t$$

$$\phi = \frac{\pi}{\theta_o} \times \theta$$

where
θ = Angle turned by cam is time t
θ = Angle turned by point R along circumference of semi-circle in time t.
ω = Angular velocity of cam
θ_0 = Angle turned by cam during rise stroke

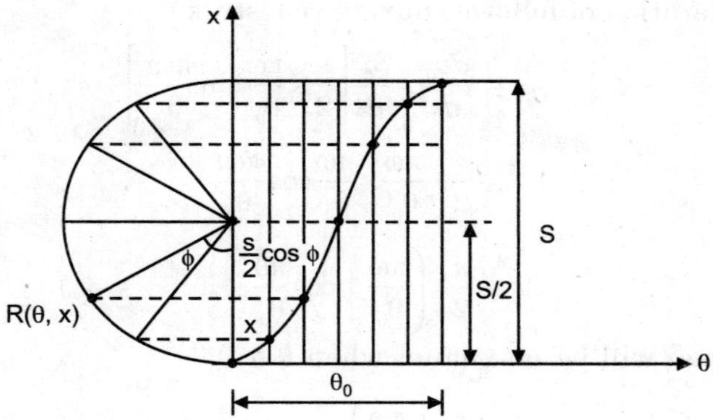

Fig. 8.11 Simple Harmonic Motion

(*i*) Displacement of follower in times *t*,

$$x = \frac{s}{2} - \frac{s}{2}\cos\phi$$

where s = stroke of the follower

$$\therefore \quad x = \frac{s}{2}(1-\cos\phi)$$

$$= \frac{s}{2}\left[1-\cos\frac{\pi\theta}{\theta_0}\right]$$

$$= \frac{s}{2}\left[1-\cos\frac{\pi\omega t}{\theta_0}\right]$$

(*ii*) Velocity of follower during rise stroke,

$$v_o = \frac{dx}{dt} = \frac{d}{dt}\left[\frac{s}{2}\left(1-\cos\frac{\pi\times\omega\times t}{\theta_0}\right)\right]$$

$$= \frac{s}{2}\left[0-\left(-\sin\frac{\pi\omega t}{\theta_0}\right)\left(\frac{\pi\omega}{\theta_0}\right)\right]$$

$$= \frac{s\times\pi\times\omega}{2\theta_0}\times\frac{\sin\pi\omega t}{\theta_0}$$

$$= \frac{s}{2}\times\frac{\pi\omega}{\theta_0}\times\sin\left(\frac{\pi\theta}{\theta_0}\right)$$

The velocity will be maximum,

When, $$\sin\frac{\pi\theta}{\theta_0} = \sin\frac{\pi}{2},$$

or $$\frac{\pi\theta}{\theta_0} = \frac{\pi}{2} \text{ and } \theta = \frac{\theta_0}{2}$$

$$(v_o)_{\max} = \frac{s}{2}\times\frac{\pi\omega}{\theta_0}\times\sin\left(\frac{\pi}{\theta_0}\times\frac{\theta_0}{2}\right) = \frac{s}{2}\times\frac{\pi\times\omega}{\theta_0}.$$

(*iii*) Acceleration of follower during rise stroke.

$$a_0 = \frac{dv_0}{dt} = \frac{d}{dt}\left[\frac{s}{2}\times\frac{\pi\omega}{\theta_0}\sin\frac{\pi\omega t}{\theta_0}\right]$$

$$= \frac{s}{2}\times\frac{\pi\omega}{\theta_0}\times\frac{\pi\omega}{\theta_0}\cos\frac{\pi\omega t}{\theta_0}$$

$$= \frac{s}{2}\times\left(\frac{\pi\omega}{\theta_0}\right)^2\cos\frac{\pi\theta}{\theta_0}$$

Acceleration will be maximum when $\theta = 0$.

$$\therefore \qquad (a_0)_{\max} = \frac{s}{2}\times\left(\frac{\pi\omega}{\theta_0}\right)^2$$

Similar analysis will give maximum velocity and acceleration during return stroke as follows.

$$(v_r)_{\max} = \frac{s}{2}\times\frac{\pi\omega}{\theta_r}$$

$$(a_r)_{\max} = \frac{s}{2}\left(\frac{\pi\omega}{\theta_r}\right)^2$$

Note : Normally $\theta_r < \theta_0$

$\therefore \qquad (v_r)_{\max} > (v_o)_{\max}$

$\qquad (a_r)_{\max} > (\alpha_o)_{\max}$

8.6.5 Parabolic Motion

Parabolic motion leads to uniform acceleration and uniform retardation of the follower. The displacement, velocity and acceleration diagram are drawn as follows.

1. Displacement Diagram

1. Divide angular displacement during rise stroke into six parts as 1, 2, 3, 4, 5, 6. Draw vertical line.
2. Divide line 3 (middle line) into six equal parts as 7, 8, 9, 10, 11, 12.
3. Join *A*-7, *A*-8, *A*-9 cutting vertical lines at point *a*, *b*, c.
4. Join *F*-10, *F*-11, *F*-12 cutting vertical lines at parts *d*, *e*, *f*.
5. Join points *A*, *a*, *b*, *c*, *d*, *f*, *F* by a smooth curve to get displacement curve for rise stroke.
6. Repeat the procedure for return stroke.

8.6.6 Analysis of Parabolic Motion

The equation of motion of follower for parabolic motion

$$s = \frac{1}{2}ut + \frac{1}{2}at^2$$

where $\qquad u = 0$ (initial velocity)

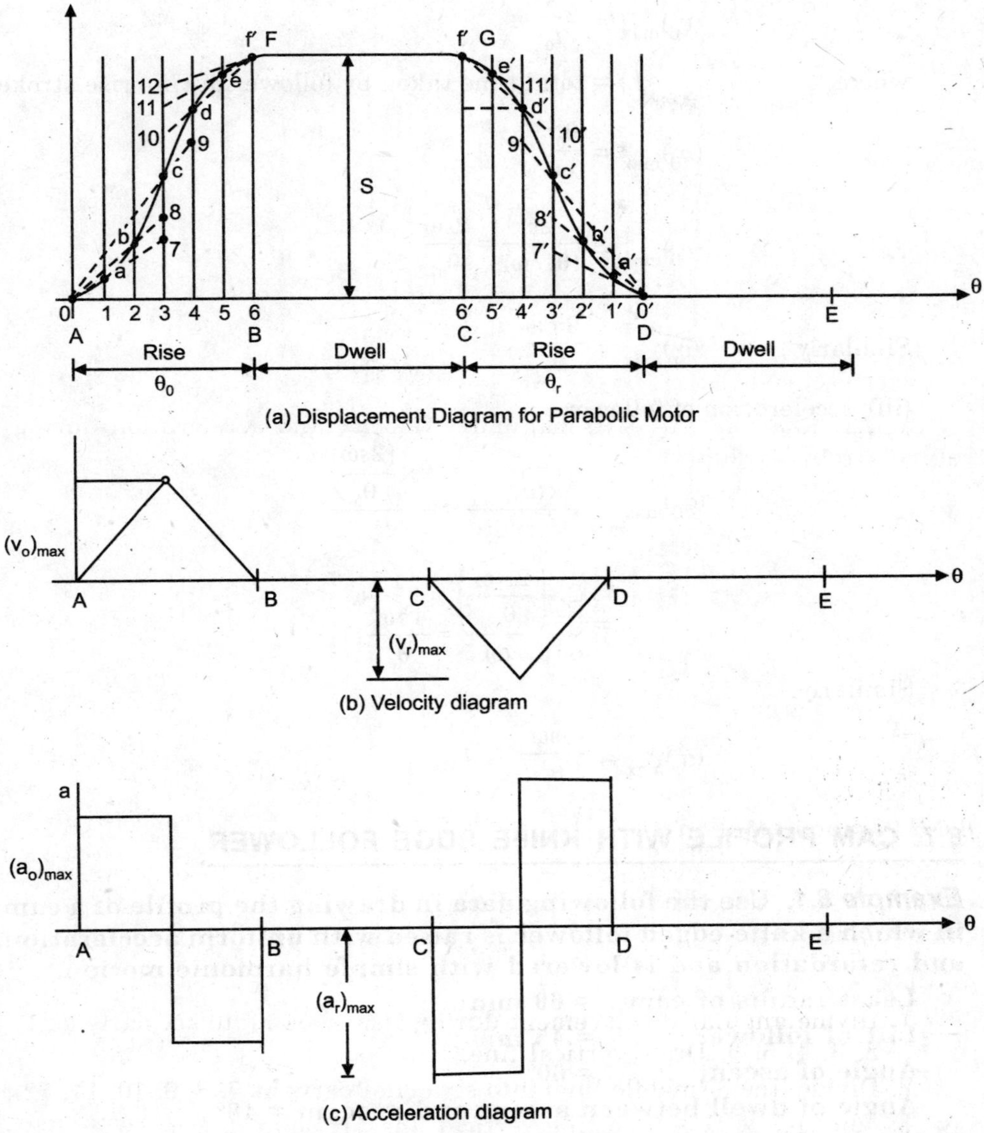

(a) Displacement Diagram for Parabolic Motor

(b) Velocity diagram

(c) Acceleration diagram

Fig. 8.12. Parabolic Motion

(*i*) $$s = \frac{1}{2}at^2$$

(*ii*) Velocity of follower

$$v = \frac{ds}{dt} = \frac{d}{dt}\left(\frac{1}{2}at^2\right) = \frac{1}{2}a^{(2t)}$$

$$= at \qquad (\because a = \text{constant})$$

$$(v_0)_{max} = 2 \times \text{Mean velocity of follower}$$

$$(v_0)_{max} = 2\frac{s}{t_0}$$

where t_0 = total time taken by follower during rise stroke

$$(v_0)_{max} = \frac{\theta_0}{\omega}$$

$$\therefore \quad (v_0)_{max} = \frac{2s}{(\theta_0/\omega)} = \frac{2s\omega}{\theta_0}$$

Similarly $(v_r)_{max} = \frac{2s\,\omega}{\theta_r}$

(iii) Acceleration of follower

$$(a_0)_{max} = \frac{2\times(v_0)_{max}}{t_0} = \frac{2\times\frac{(2s\omega)}{\theta_0}}{t_0}$$

$$= \frac{2\times\left(\frac{2s\,\omega}{\theta_0}\right)}{(\theta_0/\omega)} = \frac{4s\,\omega^2}{\theta_0^2}$$

Similarly,

$$(a_r)_{max} = \frac{4\,s\omega^2}{\theta_r^2}$$

8.7. CAM PROFILE WITH KNIFE EDGE FOLLOWER

Example 8.1. **Use the following data in drawing the profile of a cam in which a knife-edged follower is raised with uniform acceleration and retardation and is lowered with simple harmonic motion.**

Least radius of cam = 60 mm
Lift of follower = 45 mm
Angle of ascent = 60°
Angle of dwell between ascent and descent = 48°
Angle of descent = 75°.

If the cam rotates at 180 rpm, determine the maximum velocity and acceleration during ascent and descent. (UPTU 2010-11)

Solution:

I. Displacement diagram for Ascent

Draw the displacement diagram for ascent with the following data.

Angle of ascent = 60°
Lift of follower = 45 mm
Least radius of cam = 60 mm
Circumference of base circle = 2π × 60 = 377 mm.

Length of ascent stroke = $\frac{60°}{360°} \times 377 = 62.8$ mm

1. Draw a horizontal line AB = 62.8 mm. Divide into six parts 0, 1, 2, 3, 4, 5, 6. Draw vertical lines.
2. Draw a vertical line 3-*N* = 45 mm, Divide this line into six parts 3, *I, J, K, L, M, N*.
3. Join 0-*I* cutting vertical line 1 in *a*. Join 0-*J* cutting vertical line 2 in *b*. Join *f-M* cutting vertical line 5 in *e*. Join *f-L* cutting vertical line 4 in *d*. Take *K* as *c*. Join 0-*a-b-c-d-e-f* by a smooth curve to get the displacement diagram.

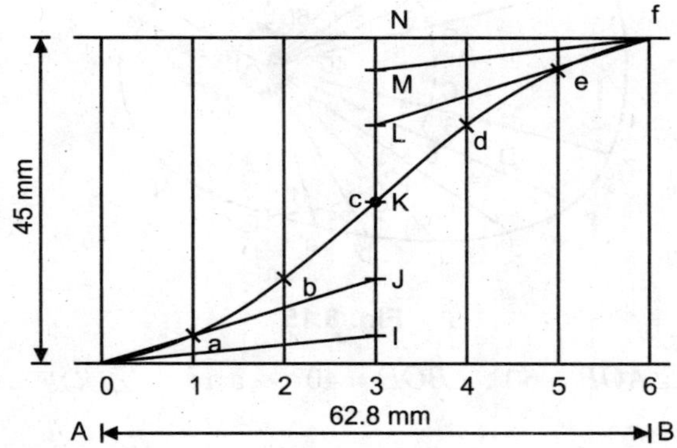

Fig. 8.13

II. Displacement Diagram for Descent

Draw the displacement diagram for descent with the following data.

Angle of descent = 75°

Length of descent stroke = $\frac{75°}{360°} \times 377 = 78.5$ mm

1. Draw a horizontal line *DE* = 78.5 mm and a vertical line EF = 45 mm. Divide *DE* into six equal parts 6', 5', 4', 3', 2', 1', 0'.

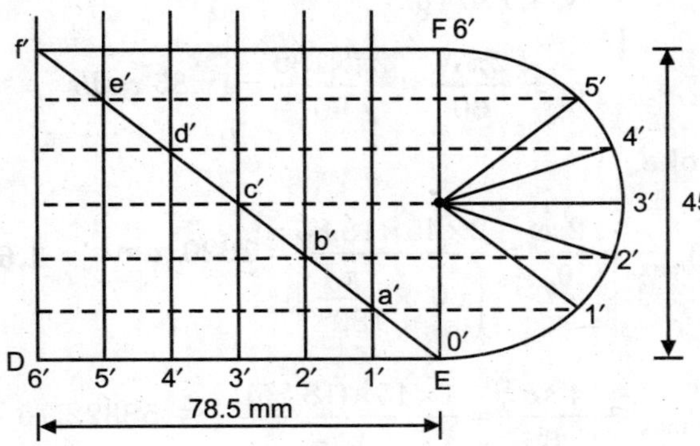

Fig. 8.14

2. Draw a semi-circle on vertical line EF and divide into six equal arcs 1' 2' 3' 4' 5' 6'. Draw horizontal lines through these points cutting the vertices lines in points *f' e' d' c' b' a'* 0'. Join these points by a smooth curve to get the displacement diagram.

III. Cam Profile Construction

1. Draw base circle with centre O and radius 60 mm.

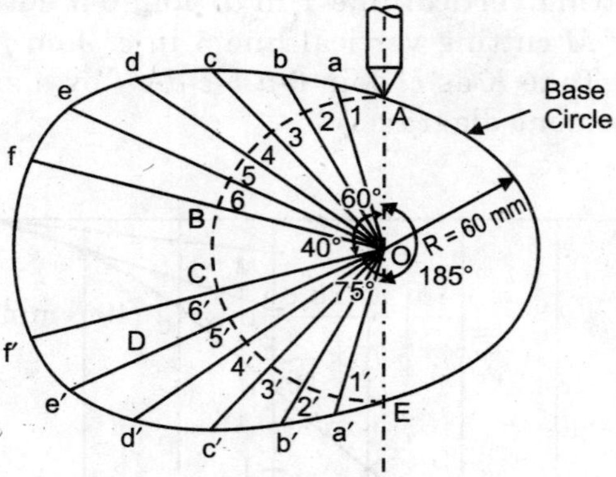

Fig. 8.15

2. Draw $\angle AOB = 60°, \angle BOD = 40°$ and $\angle DOE = 75°$ so that $\angle EOA = 185°$.
3. Divide $\angle AOB$ into six equal parts and extend radial lines. *O*-1, *O*-2, *O*-3, *O*-4, *O*-5, *O*-6.
4. Divide $\angle DOE$ into six equal parts and extend radial lines *O*-1', *O*-2', *O*-3', *O*-4' *O*-5' and *O*-6'.
5. From displacement diagram cut radial lines equal to 1-*a*, 2-*b*, 3-*c*, 4-*d*, 5-*e*, 6-*f* and 6'-*f* 5'-*e'*,4'-*d'*, 3'-*c'*,2'-*b'*, 1'-*a'*.
6. Join points *A*, *a*, *b*, *c*, *d*, *e*, *f*, *f'*, *e'*, *d'*, *c'*, *b'*, *a'*, *E* and *A* by smooth curve to get the cam profile.

IV. Calculations

$$N = 180 \text{ rpm}$$

$$\omega = \frac{2\pi N}{60} = \frac{2\pi \times 180}{60} = 18.85 \text{ rad/s}$$

For ascent stroke,

$$(v_0)_{max} = \frac{2s\omega}{\theta_0} = \frac{2 \times 45 \times 18.85}{\left(60° \times \frac{\pi}{180°}\right)} = 1620 \text{ mm/s} = \mathbf{1.62 \text{ m/s Ans.}}$$

$$(\alpha_0)_{max} = \frac{4\,s\omega^2}{\theta_0^2} = \frac{4 \times 45 \times (18.85)^2}{\left(60 \times \frac{\pi}{180}\right)^2} = 58322.75 \text{ mm/s}^2$$

$$= 58.32 \text{ m/s}^2 \quad \textbf{Ans.}$$

For descent stroke

$$(v_r)_{max} = \frac{s}{2}\times\frac{\pi\omega}{\theta_r}$$

$$= \frac{\frac{45}{2}\times\pi\times 18.85}{75\times\frac{\pi}{180}} = 1017.5 \text{ mm/s}$$

$= \mathbf{1.0175\ m/s}$ **Ans.**

$$(a_r)_{max} = \frac{s}{2}\left(\frac{\pi\omega}{\theta_r}\right)^2$$

$$= \frac{45}{2}\left(\frac{\pi\times 18.85}{75\times\frac{\pi}{180}}\right)^2 = 46049.8 \text{ mm/s}^2$$

$= \mathbf{46.05\ m/s^2}$ **Ans.**

8.8. CAM PROFILE WITH RADIAL ROLLER FOLLOWER

A roller follower is used in place of knife-edge-follower to reduce its friction and wear. The path of the roller is determined by the same method as for knife-edged follower.

Example 8.2. **Draw the profile of a cam to raise a valve with harmonic motion through 50 mm in 1/3 of revolution, keep it fully raised through 1/12 revolution, and to lower it with harmonic motion in 1/16 revolution. The valve remains closed during rest of the cam rotation. The minimum radius of cam is 25 mm and radius of roller is 10 mm and diameter of cam shaft is 25 mm. The axis of the valve shaft passes through the axis of the cam shaft. The cam shaft is rotating with uniform velocity.**

Solution: Lift, $S = 50$ mm

Outstroke angle, $\theta_0 = 1/3 \times 360° = 120°$.

Angle of dlwell after ascent $= \frac{1}{12}\times 360° = 30°$

Return stroke angle, $\theta_r = \frac{1}{6}\times 360° = 60°$

Angle of dwell after return stroke

$= 360° - (120° + 30° + 60°) = 150°$.

Radius of roller $= 10$ mm

Minimum radius of cam $= R_{min} = 25$ mm

Cam shaft diameter $= 25$ mm

I. Displacement Diagram

1. Select a suitable horizontal and vertical scales.
2. Draw a horizontal line $AB = 360°$.
   Cut $AC = 120°$, $CD = 30°$ and $DE = 60°$, $EB = 150°$.
3. Draw vertical line $AF = 50$ mm.
4. Draw semi-circle on AF and divide it into eight parts. Join 1, 2, 3, 4, 5, 6, 7, 8 with centre of semi-circle.
5. Through above points draw horizontal lines.
6. Divide horizontal lines AC and DE into 8 parts each. Draw vertical lines.
7. The vertical lines and horizontal lines intersect at, a, b, c, d, e, f, g h, and h', g', f', e', d', c', b', a'.
8. Join points A, a, b, c, d, e, f, g, h, h', g', f', e', d', c', b', a', E with a smooth cannot to get displacement diagram.

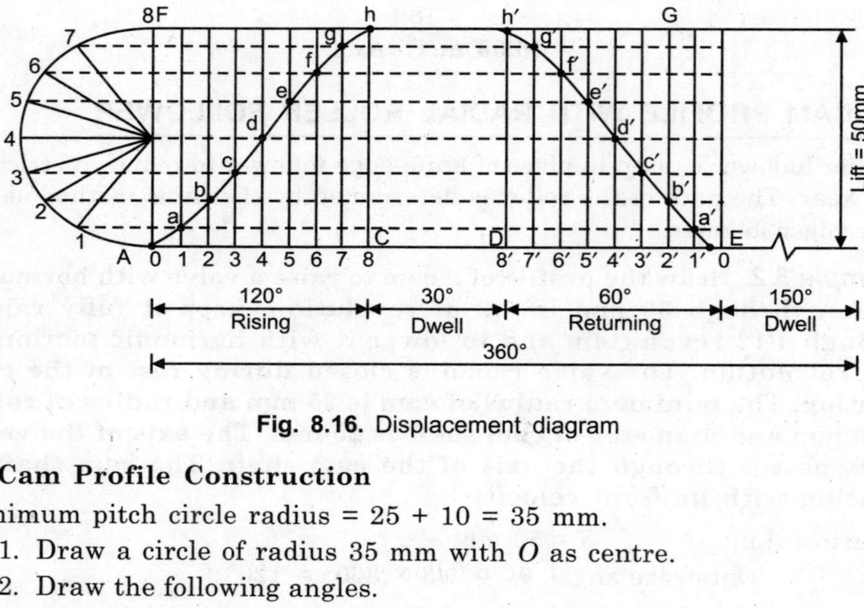

Fig. 8.16. Displacement diagram

II. Cam Profile Construction

Minimum pitch circle radius = 25 + 10 = 35 mm.

1. Draw a circle of radius 35 mm with O as centre.
2. Draw the following angles.

$$\angle AOC = 120°, \angle COD = 30°, \angle DOE = 60°, \angle EOA = 150°.$$

3. Divide arc AC and DE into eight equal parts each and mark 1, 2, 3, 4, 5, 6, 7, 8 and 8', 7', 6', 5',4', 3', 2', 1'.
4. Join the points with centre O and extend upto prime circle.
5. On radial lines mark the distances from displacement diagram.
6. With points a, b, c, d, e, f, g, h, h', g', f', e', d', c', b', a', as centre and radius = 10 mm draw circles.
7. Join the bottom of these circles with a smooth curve to get the cam profile.

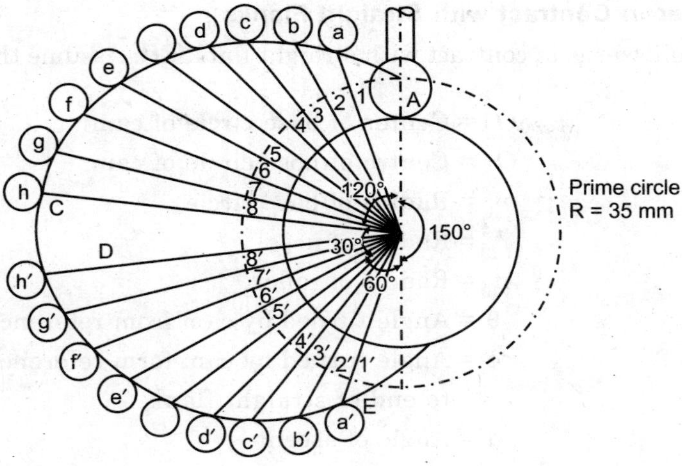

Fig. 8.17. Cam Profile

8.9. ANALYTICAL METHODS OF CAM DESIGN

The graphical method of cam profile construction is simple and quick. However, the accuracy achieved is low and may not be acceptable in many cases specially high speed cams. The availability of computers has made the analytical methods very attractive. The analytical method results in higher accuracy and optimal design.

The follower displacement (x) is expressed as a function of cam rotation (θ) and cam profile is analytically obtained in the form of parametric equations. The basic principle is similar to graphical method. The radical distance for cam rotation angle is obtained.

The shape of the cam profile is built of circular arcs and straight lines. The cam can thus be manufactured accurately, easily and economically. Such cams are called cams with specified contours. The following are important cams of specified contours.

1. Tangent cam with roller follower.
2. Circular cam with flat faced follower.

8.10. TANGENT CAM WITH ROLLER FOLLOWER

A tangent cam has symmetrical profile about the centre line and consists of two straight flanks tangent to base circle and nose circle. This type of cam is widely used for operating the inlet and exhaust valves of internal combustion engines. The cam profile consists of two straight lines and two circular arcs, the cam surface can be accurately machined.

In order to find displacement, velocity and acceleration of the roller follower, two positions of the roller may be analysed.

1. The roller in contact with straight flanks.
2. The roller in contact with nose circle.

8.10.1 Roller in Contract with Straight Flanks

The roller follower is in contract with straight flank EG. Assume the following parameters.

O_1 = Center of base circle of cam

O_2 = Centre of nose circle of cam

r_1 = Radius of base circle

r_2 = Radius of nose

r_3 = Radius of roller

θ = Angle turned by cam from reference position

ϕ = Angle turned by cam form reference position to end of straight flank.

α = Angle of ascent

ω = Angular velocity of cam.

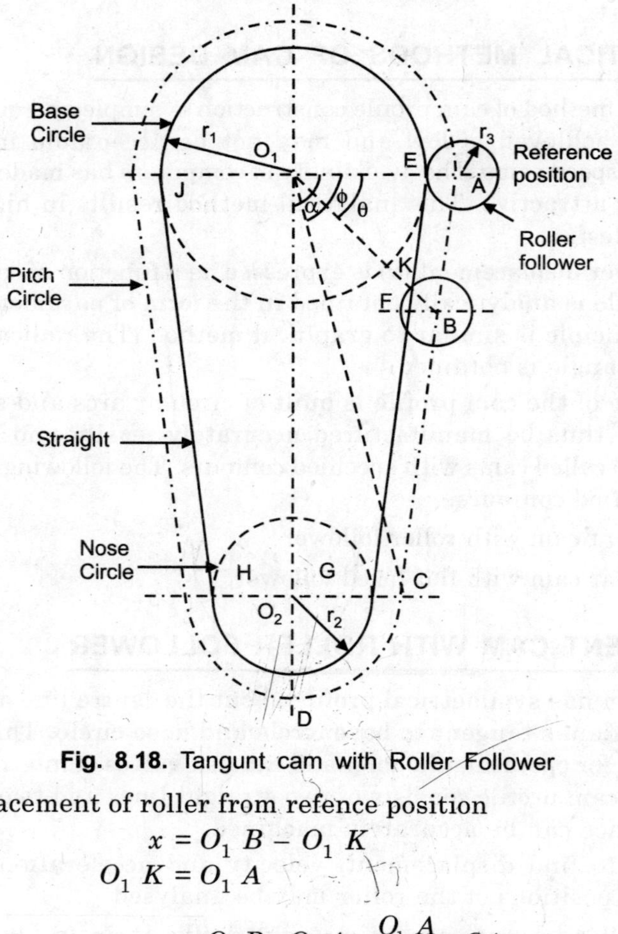

Fig. 8.18. Tangunt cam with Roller Follower

The displacement of roller from refence position

$$x = O_1 B - O_1 K$$

$$O_1 K = O_1 A$$

$$\therefore \quad x = O_1 B - O_1 A = \frac{O_1 A}{\cos\theta} - OA$$

$$\because \quad \text{From } \Delta O_1 - AB, \; \frac{O_1A}{O_1B} = \cos\theta$$

$$\therefore \quad O_1B = \frac{O_1A}{\cos\theta}$$

$$\therefore \; x = O_1 A\left(\frac{1}{\cos\theta} - 1\right) = O_1A\left(\frac{1-\cos\theta}{\cos\theta}\right).$$

$$a_{min} = \omega^2(r_1 + r_3)\; \frac{(2-1)}{1}$$

$$= \omega^2(r_1 + r_3)$$

8.10.2 The Roller in Contact with Nose Circle

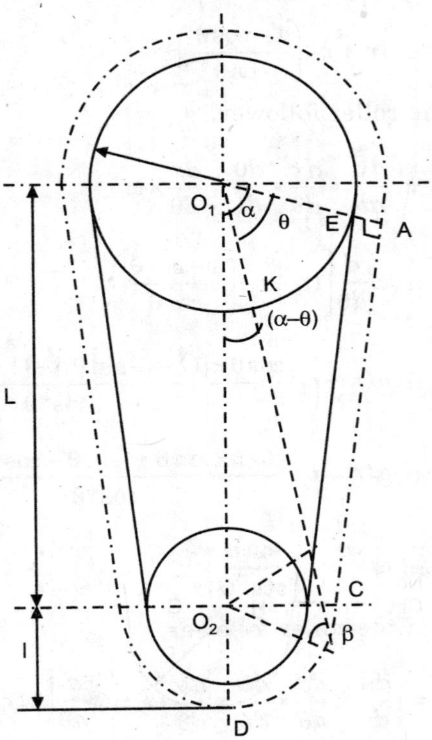

Fig. 8.19. Roller follomer in contract with nose of Tangent cam

The displacement

$$x = O_1\, B - O_1K$$
$$= O_1\, B - O_1A.$$

From geometry

$$x = L\cos(\alpha-\theta) + \sqrt{l^2 - L^2\sin^2(\alpha-\theta)]} - (r_1 + r_3)$$

The velocity,

$$v = \frac{dx}{dt}$$

$$= \omega \times L \left| \sin(\alpha - \theta) + \frac{L \sin 2(\alpha - \theta)}{2\left[l^2 - L^2 \sin^2(\alpha - \theta)\right]^{3/2}} \right.$$

The acceleration,

$$\alpha = \frac{dv}{dt}$$

$$= -\omega^2 \times L \left| \cos(\alpha - \theta) + \frac{l^2 \times L \cos_2(\alpha - \theta) - L^3 \sin^4(\alpha - \theta)}{2[l^2 - L^2 \sin^2(\alpha - \theta)]^{3/2}} \right|$$

Now, $O_1A = O_1E + EA = r_1 + r_3$.

$$\therefore \quad x = (r_1 + r_3)\left(\frac{1 - \cos\theta}{\cos\theta}\right) \quad \text{...(1)}$$

The velocity of the roller follower,

$$v = \frac{dx}{dt} = \frac{dx}{d\theta} \times \frac{d\theta}{dt} = \frac{dx}{d\theta} \times \omega$$

$$= \omega \frac{d}{d\theta}\left[(r_1 + r_3)\left(\frac{1 - \cos\theta}{\cos\theta}\right)\right]$$

$$= \omega(r_1 + r_3)\frac{\cos\theta \times [O - (-\sin\theta)] - (1 - \cos\theta)(-\sin\theta)}{\cos^2\theta}$$

$$= \omega(r_1 + r_3)\frac{\cos\theta \times \sin\theta + (\sin\theta - \cos\theta\sin\theta)}{\cos^2\theta}$$

$$= \omega(r_1 + r_3)\frac{\sin\theta}{\cos^2\theta} \quad \text{...(2)}$$

The acceleration of the roller follower,

$$a = \frac{dv}{dt} = \frac{dv}{d\theta} \times \frac{d\theta}{dt} = \frac{dv}{d\theta}.\omega = \omega\frac{d}{d\theta}\left[\omega(r_1 + r_3)\frac{\sin\theta}{\cos^2\theta}\right]$$

$$= \omega^2(r_1 + r_3)\frac{\cos^2\theta(\cos\theta) - \sin\theta(2\cos\theta)(-\sin\theta)}{\cos^4\theta}$$

$$= \omega^2(r_1 + r_3)\frac{\cos^2\theta(\cos^2\theta + 2\sin^2\theta)}{\cos^4\theta}$$

$$= \omega^2(r_1 + r_3)\frac{2 - \cos^2\theta}{\cos^3\theta} \quad \text{... (3)}$$

From equation (2), velocity will be maximum when $\frac{\sin\theta}{\cos^2\theta}$ is maximum or where $\theta = \phi$

$$\therefore \quad v_{max} = \omega(r_1 + r_3)\frac{\sin\phi}{\cos^2\phi}$$

From equation (3), acceleration will be maximum when $\frac{2-\cos^2\theta}{\cos^2\theta}$ is maximum, *i.e.*, when $\theta = \phi$.

$$\therefore \quad a_{max} = \omega^2(r_1 + r_3)\left(\frac{2-\cos^2\phi}{\cos^3\phi}\right).$$

The acceleration is minimum when $\theta = 0$.

Example 8.3. **A tangent cam with straight working faces tangential to a base circle of 120 mm diameter has a roller diameter 48 mm. The lime of stroke of the roller follower passes through the axis of the cam. The nose circle radius of the cam is 12 mm and the angle between the tangential faces of the cam is 90°. If the speed of the cam is 180 rpm, determine the acceleration of the follower when.**

(*i*) During the lift, the roller just leaves the straight flank.

(*ii*) The roller is at the outer end of its lift, *i.e.*, at the top of the nose. [UPTU 2010-11]

Solution:

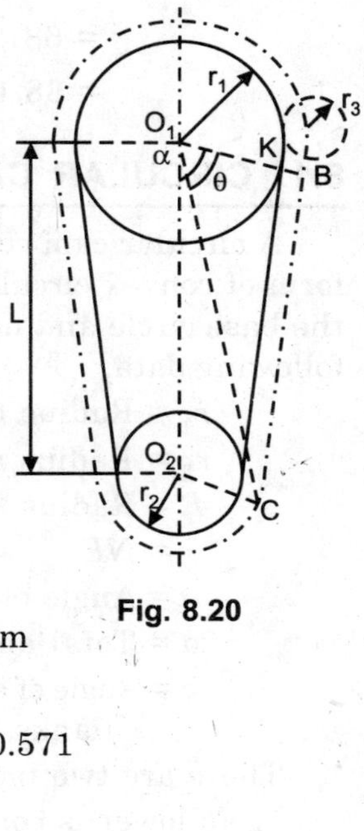

Fig. 8.20

Radius of base circle, $r_1 = \frac{120}{2} = 60$ mm

Radius of nose circle, $r_2 = 12$ mm

Radius of the roller, $r_3 = \frac{48}{2} = 24$ mm

The angular velocity, $\omega = \frac{2\pi N}{60}$

$$= \frac{2\pi\times180}{60} = 18.65 \text{ rad/s}$$

$$\alpha = 180° - 45° - 90° = 45°$$

$$r_1 = L\cos\alpha + r_2$$

$$L = \frac{r_1 - r_2}{\cos\alpha} = \frac{60-12}{\cos 45°} = 67.88 \text{ mm}$$

$$\tan\theta = \frac{L\sin\alpha}{r_1 + r_3} = \frac{67.88\sin 45°}{60+24} = 0.571$$

$$\therefore \quad \theta = 29.74°.$$

(*i*) Acceleration when the roller just leaves the straight flank.

$$\alpha = \frac{\omega^2 (r_1 + r_3)(2 - \cos^2\theta)}{\cos^3\theta}$$

$$= \frac{(18.65)^2 (60 + 24)(2 - \cos^2 29.74°)}{\cos^3 29.74°}$$

$$= \frac{347.82 \times 84 \times 1.246}{0.6546}$$

$$= 55613 \text{ mm/s}^2$$

$$= \mathbf{55.613 \text{ m/s}^2} \quad \textbf{Ans.}$$

(*ii*) Acceleration when the roller is at the outer end of its lift, *i.e.,* at the tep of the nose.

$$\theta = \alpha.$$

$$\alpha = \omega^2 L\left[-\cos(\alpha-\theta) - \frac{L^3 \sin^2 2(\alpha-\theta)}{4[l^2 - L^2)(\alpha-\theta)]^{3/2}} - \frac{L\cos 2(\alpha-\theta)}{l^2 - L^2 \sin(\alpha-\theta)}\right]$$

$$= \omega^2 L\left[-1 - \frac{L}{l}\right]$$

$$l = r_2 + r_3 = 24 + 12 = 36 \text{ mm.}$$

$$\alpha = (18.65)^2 67.88\left[-1 - \frac{67.88}{36}\right]$$

$$= 6812.85 \text{ mm/s}^2$$

$$= \mathbf{68.1285 \text{ m/s}^2} \quad \textbf{Ans.}$$

8.11 CIRCULAR CAM WITH FLAT FACED FOLLOWER

A circular cam consists of flanks in the form of convex circular arcs which connect the base circle and nose circle. Assume the following data:

r_1 = Radius of base circle = O_1E
r_2 = Radius of nose circle = O_2L
R = Radius of circular flanks = NE
α = Angle of ascent
2α = Total angle of action of cam.
ϕ = Angle of action of cam on circular flanks.

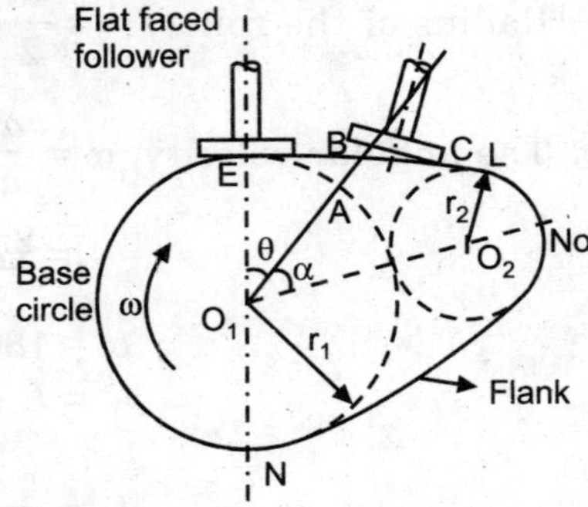

Fig. 8.21 Circular cam with flat-faced follower.

There are two important cases for analysis.

1. Follower is contact with circular flanks

2. Follower on the nose.

8.11.1 Follower on Circular Flanks

$$x = AB = BO_1 - AO_1$$

Applying geometry

$$x = (R - r_1)(1 - \cos\theta) \qquad ...(1)$$

where θ = angular displacement of cam.

Velocity of the follower,

$$v = \frac{dx}{dt} = \frac{dx}{d\theta} \times \frac{d\theta}{dt} = \omega \times \frac{d}{d\theta}(R - r_1)(1 - \cos\theta)$$

$$= \omega(R - r_1)\sin\theta \qquad ...(2)$$

Acceleration of the follower

$$\alpha = \frac{dv}{dt} = \frac{dv}{d\theta} \times \frac{d\theta}{dt} = \omega \times \frac{d}{d\theta}(\omega(R - r_1)\sin\theta)$$

$$= \omega^2 \times (R - r_1)\cos\theta$$

Velocity is maximum when $\theta = \phi$.

$$v_{max} = \omega(R - r_1)\sin\phi$$

Acceleration is maximum when $\theta = 0$ and $\cos\theta = 1$

$$a_{max} = \omega(R - r_1).$$

8.11.2 Follower on the Nose

The displacement,

$$x = AB - BO_1 - AO_1$$

Using geometry,

$$x = CO_2 + O_2O_1 \cos(\alpha - \theta) - AO_1$$

Velocity of the follower

$$v = \frac{dx}{dt} = \frac{dx}{d\theta} \times \frac{d\theta}{dt} = \omega \times \frac{d}{d\theta}[CO_2 + O2O_1 \cos(\alpha - \theta) - AO_1]$$

$$= \omega \times O_1O_2 \sin(\alpha - \theta)$$

The acceleration of the follower

$$\alpha = \frac{dv}{dt} = \frac{dv}{dt} \times \frac{d\theta}{dt} = \omega \times \frac{d}{d\theta}[\omega \times O_1O_2 \sin(\alpha - \theta)]$$

$$= \omega^2 O_1O_2 \cos(\alpha - \theta)$$

The retardation is max. where $(\alpha - \theta) = 0$ or $\theta = \alpha$.

$$\therefore \quad a_{max} = -\omega^2 O_1O_2.$$

where O_1O_2 = Distance between cam centre and nose centre.

Example 8.4. **A symmetrical circular cam with flat faced follower has least radius of 27.5 mm, nose radius of 3 mm, total lift is 12.5 mm, angle of lift is 55°. Find the following if cam speed is 600° rpm.**

(*a*) Distance between cam centre and nose centre

(*b*) Radius of circular flanks.

(*c*) Angle of contact on the circular flank.

Solution :

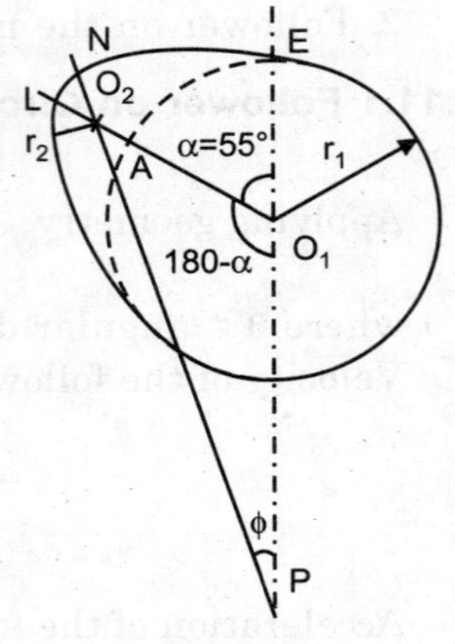

Fig. 8.22

$$r_1 = 30 \text{ mm}$$
$$\text{Lift} = 12.5 \text{ mm}$$
$$r_2 = 3 \text{ mm}$$
$$\alpha = 55°$$
$$N = 600 \text{ rpm}$$

Assume $PE = R$ = Radius of circular flank.

O_1O_2 = Distance between cam centre and nose centre.

$\phi = \angle O_1 PO_2$ = Angle of contact on the circular flank.

AL = Total lift of the cam.

(a) Distance between cam centre and nose centre

$$AL = O_1L - O_1A$$
$$= OL - r_1$$
$$= (O_1O_2 + LO_2) - r_1$$
$$= O_1O_2 + r_2 - r_1$$
$$AL = \text{Total lift} = 12.5 \text{ mm}$$
$$\therefore \quad O_1O_2 = AL - r_2 + r_1$$
$$= 12.5 - 3 + 30 = 39.5 \text{ mm} \quad \textbf{Ans.}$$

(b) Radius of circular flank

From figure

$$PN = PE = R. = PO_2 + O_2N = PO_2 + r_2$$

Now
$$PO_2 = PN - r_2 = R - r_2 = R - 3$$
$$PE = PO_1 + O_1E$$
$$R = PO_1 + r_1$$
$$PO_1 = R - r_1 = R - 30.$$

From $\Delta O_1 PO_2$

$$(PO_2)^2 = \Delta O_1 PO_2 + (O_1 O_2)^2 - 2O_1P \times O_1O_2 \cos(180° - \alpha)$$
$$(R - 3)^2 = (R - 30)^2 + 39.5^2 - 2 \times (R - 30)(39.5)\cos(180°–55°)$$
$$R = 125.54 \text{ mm} \quad \textbf{Ans.}$$

(c) Angle of contact on circular flank

Using Lami's Theorem for $\Delta O_1 PO_2$

$$\frac{PO_2}{\sin(180° - \alpha)} = \frac{O_1O_2}{\sin\phi}$$

$$\therefore \quad \sin\phi = \frac{O_1O_2 \sin(180° - \alpha)}{PO_2} = \frac{39.5 \sin(180° - 55°)}{(R - r_2)}$$

$$= \frac{39.5 \sin 125°}{(125.54 - 3)} = 0.264$$

$$\phi = 15.31° \quad \textbf{Ans.}$$

QUESTION BANK

1. Give a detailed classification of cams and followers.
2. Define the following:
   (*i*) Base circle
   (*ii*) Pitch circle
   (*iii*) Trace circle
   (*iv*) Pressure angle
3. Deduce the expression for the velocity and acceleration of the follower when it moves with simple harmonic motion. Why simple harmonic motion programme is not suitable for high speed cams? **[UPTU, 2010-11]**
4. Define the following:
   (*a*) Kinematic derivatives
   (*b*) Under-cutting of cam.
   (*c*) Analytical method of cam designs.
5. Differentiate between a tangent cam and a circular cam. Compare their advantages and disadvantages.
6. Explain the procedure to layout a cam profile graphically.

NUMERICAL PROBLEMS

1. Draw the profile of a cam that gives a lift of 40 mm to a roller follower of 20 mm diameter. The axis of the follower passes through the centre of the cam. The base circle radius of cam is 50 mm. The follower is to be lifted with simple harmonic motion in a quarter revolution and to be dropped suddenly at half revolution. Determine the maximum velocity and max. acceleration during the lifting. The cam rotates at 60 rpm. [**Ans.** 0.25 m/s' 3.155 m/s^2]
2. A flat-faced follower is operated by a circular cam. The total action is 150° angle, base diameter 125 mm and lift is 25 mm. During the lift, the period of acceleration is half that of decleration, speed of cam shaft is 1250 rpm. The straight line of tappet passes through the cam axis. Find
   (*a*) Radii of the none and flank
   (*b*) Maximum acceleration and deceleration during the lift.
   [**Ans.** (*a*) 40.3 mm, 148 mm, (*b*) 1465 m/s^2, 808.8 m/s^2]

UNIT–V

9. GEARS

10. GEAR TRAINS

CHAPTER 9

GEARS

9.1. INTRODUCTION

Gears are very important machine elements mainly in the form of toothed wheels used for transmitting rotary motion from one shaft to another. The main characteristics of gears are as follows.

1. Gears are precisely machined elements.
2. These provide positive and smooth drive without noise.
3. There is no slip or creep and exact velocity ratios are obtained.
4. Gears are used to transmit large power from driver to driven machine.
5. Gears are used to transmit motion between shafts which are very closely spaced and thereby very compact mechanism and machines are possible.

In the light of the above attractions, gears and gear trains are very widely used in machine building especially automobiles, machine tools and other process machines.

9.2. CLASSIFICATION OF GEARS

A very wide variety of gears and their drive arrangements are used in the industry. Important classification factors may be as follows.

1. Position of shaft axes
2. Gear speed
3. Angle of mating gear teeth
4. Type of gearing

   Important type of gears are shown in Fig. 9.1.

9.2.1 Position of shaft Axes

(*a*) Parallel shafts

1. Spur gears with straight teeth are commonly used to connect parallel shafts. Spur gears can be external gear as well as internal gears. These gears have line contact.
2. Helical gears have inclined or curved teeth to the axes of the shafts. Each teeth is helical in shape and have point contact.
3. Double helical gears have right-handed and left-handed helix. These are also known as herring bone gears and meet at a common apex.

(*b*) Intersecting shafts

Bevel gears are used to connect intersecting shafts. The bevel teeth can be straight or inclined in the helical or spiral bevel gear.

(*c*) Skew shafts are neither parallel nor intersecting. These shafts are connected with the help of skew spiral gears and skew bevel gears.

9.2.2. Gear Speed

(*a*) Low velocity gears have speed less than 2 m/s.

(*b*) Medium velocity gears have speed between 3 m/s and 15 m/s.

(*c*) High velocity gears have speed above 15 m/s.

9.2.3. Type of Gearing

(*a*) External gears are used to connect the shafts externally.

(*b*) Internal gears are used to connect a small wheel with external teeth called a pinion with an annular wheel with internal gears.

(*c*) In rack and pinion gear a straight line gear called a rack is connected to a circular wheel called pinion. It converts a rotary motion into linear motion or vice-versa.

(a) Spur gears

(b) Helical gears

(c) Miter gears

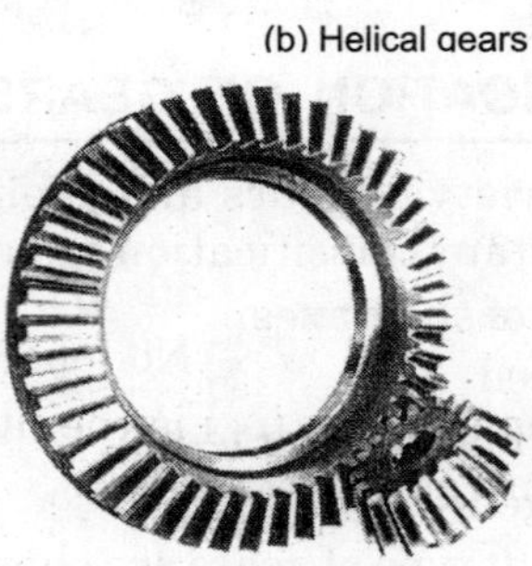

(d) Bevel gears

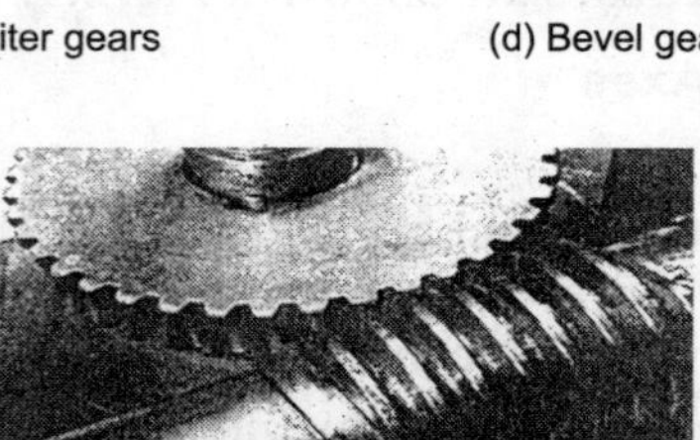

(e) Worm and worm wheel

Fig. 9.1 Types of Gears

9.3. TERMINOLOGY USED IN GEARS

The profile of a gear alongwith important terms which are used for study of gears is shown in Fig. 9.2.

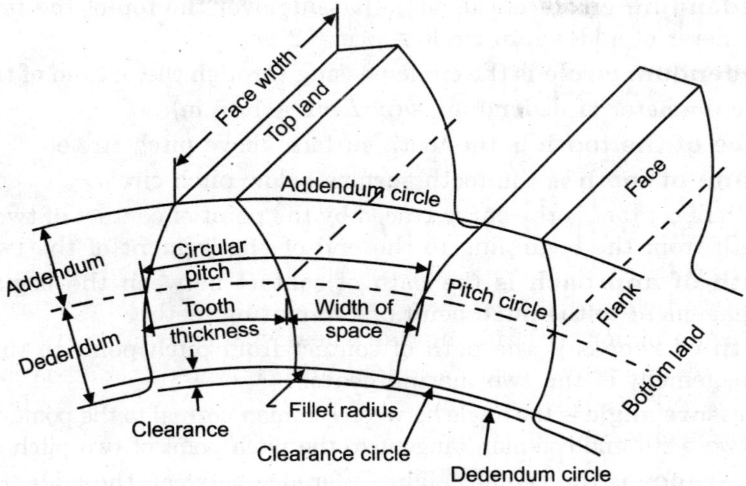

Fig. 9.2 Gear Terminology

1. **The pitch circle diameter** is the diameter of a pitch circle or a circle which would produce same rotation by pure rolling.
2. **Pitch point** is the common point of contact of pitch circles of two mating gears.
3. **Circular pitch** (p_c) is the distance between a point on one tooth to corresponding point on adjacent tooth along the circumference of the pitch circle.

   The circular pitch

$$p_c = \frac{\text{circumference of pitch circle}}{\text{No. of teeth}} = \frac{\pi D}{T}$$

where D = Pitch circle diameter

T = No. of teeth on the gear

4. **Diametrical pitch** (p_d) is the number of teeth per unit length of pitch circle diameter.

$$p_d = \frac{T}{D}$$

$$\therefore \quad p_c \times p_d = \frac{\pi D}{T} \times \frac{T}{D} = \pi$$

5. **Module** (m) is the length of pitch circle circumference per tooth

$$m = \frac{D}{T} = \frac{1}{p_d} \quad \text{[mm]}$$

$$\therefore \quad p_c = \frac{\pi}{p_d} = \frac{\pi \times D}{T} = \pi \times m.$$

6. Addendum (a) is the radial distance of the tooth, above pitch circle.

7. **Dedendum** is the radial distance of the tooth below pitch circle.

   The dedendum = 1.157 m = $\left(1+\frac{\pi}{20}\right)m$.
8. **Addendum circle** is the circle passing over the top of the teeth. The diameter of addendum circle = *pcd* + 2 *m*.
9. **Dedendum circle** is the circle passing through the bottom of the teeth. The diameter of dedendum = (*pcd* – 2 × 1.15 m)
10. **Face of the tooth** is the tooth surface above pitch circle.
11. **Flank of tooth** is the tooth surface below pitch circle.
12. Path of contact is the curve traced by the point of contact of two mating teeth from the beginning to the end of engagement of the two teeth.
13. **Path of approach** is the path of contact between the beginning of engagement to the pitch point of two mating teeth.
14. **Path of recess** is the path of contact from pitch point to the end of engagement of the two mating gears.
15. **Pressure angle** is the angle between common normal to the point of contact of two teeth and common tangent to the pitch point of two pitch circle.
16. **Clearance** is the radial height difference between the addendum and dedendum.

    Clearance = 1.15 nm – m = 0.157 m.
17. **Profile** is the curve forming face and flank of the tooth.
18. **Pinion** is the smallest and driving gear of a mated gear pair.
19. **Rack** is the straight gear or gear of infinite diameter.
20. **Gear Ratio** (*G*) is the ratio of number of teeth of gear wheel to that of pinion

    $$G = \frac{T_2}{T_1}$$

    where T_1 = No. of teeth on pinion (driver)

    T_2 = No. of teeth on gear wheel (driven)
21. **Velocity Ratio** (*VR*) is the ratio of angular velocity of the follower and driving gear respectively.

    $$\text{V.R.} = \frac{\omega_2}{\omega_1} = \frac{2\pi N_2}{2\pi N_1} = \frac{N_2}{N_1}$$

    Now $$\pi d_1 N_1 = \pi d_2 N_2$$

    $$\therefore \quad \frac{N_2}{N_1} = \frac{d_1}{d_2}$$

    $$\therefore \quad \text{V.R.} = \frac{d_1}{d_2}$$

    Now $$p_c = \frac{\pi d_1}{T_1} = \frac{\pi d_2}{T_2}$$

    $$\therefore \quad \frac{d_1}{d_2} = \frac{T_1}{T_2}$$

$$\therefore \quad V.R. = \frac{T_1}{T_2} = \frac{1}{G}$$

22. **Line of Action or pressure line** is the common normal to the point of contact of mating gears.

Example 9.1. A spur gear of 30 teeth has a module of 2 mm and pitch line velocity of 0.6283 m/s. Determine:

(a) Gear speed

(b) Circular pitch.

Solution : $T = 30$

$m = 2$ mm

$V_p = 0.6283$ m/s $= 628.3$ mm/s.

(b) Now circular pitch,

$$p_c = \pi \times m = \pi \times 2 = \mathbf{6.28\ mm\ Ans.}$$

(a) Pitch line velocity,

$$V_p = \omega \times r$$

$$= \frac{2\pi N}{60} \times \frac{D}{2}$$

But $D = m \times T$

$$\therefore \quad V_p = \frac{2\pi N}{60} \times \frac{mT}{2}$$

$$\therefore \quad N = \frac{60\times 2\times V_p}{2\pi mT} = \frac{60\times 2\times 628.3}{20\times 2\times 30} = \mathbf{200\ rpm\ Ans.}$$

9.4. LAW OF GEARING

The ratio of the angular velocities of the driven gear and driving gear must remain constant through entire phase of contact of gear teeth. This is the fundamental criterion that governs the choice of tooth profile. Otherwise very serious vibration and impact problems would arise even at low speeds.

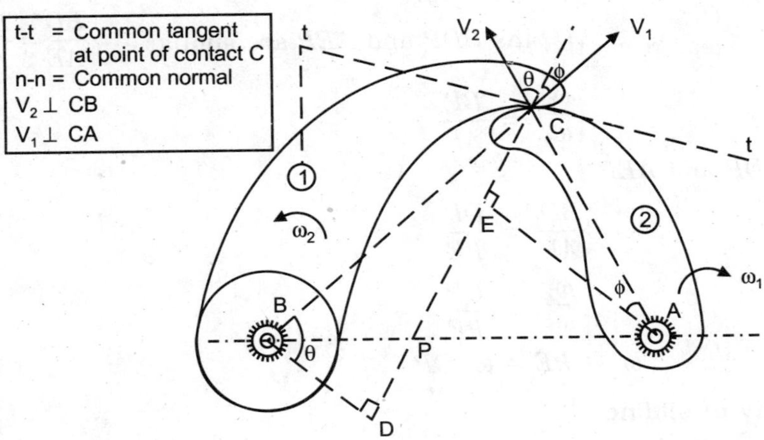

Fig. 9.3. Law of Gearing.

The above condition can be interpreted so that the common normal to point of contact of the two surfaces intersects the line joining the centre of rotation of two surfaces at a fixed point and divides the centre distance inversely as the ratio of angular velocities.

Two curved bodies 1 and 2 rotate about fixed centres A and B respectively with point of contact at C.

V_1 = Linear velocity of point C when on body 1

V_2 = Linear velocity of point C when on body 2

P = Point of contact of common normal $(n - n)$ and centre line AB.

ω_1 = Angular velocity of body 1

ω_2 = Angular velocity of body 2

θ = Angle between V_2 and common normal $n - n$

ϕ = Angle between V_1 and common normal $n - n$

$$\angle BCV_2 = 90°$$

$$\text{Angle } DCB = 180° - \angle BCV_2 - \theta = 180° - 90° - \theta = 90° - \theta$$

$$\text{Angle } ACE = 180 - \phi - \angle ACV_1 = 180° - \phi - 90° - \phi$$

Component of V_1 along $n - n = V_1 \cos \phi$

Component of V_2 along $n - n = V_2 \cos \theta$

$\therefore$ Relative motion along $n - n = V_1 \cos \phi - V_2 \cos \theta$

Relative motion along $n - n = 0$ [For proper contact]

$$\therefore \quad V_1 \cos \phi - V_1 \cos \theta = 0$$

$$V_1 = \omega_1 AC$$

$$V_2 = \omega_2 BC$$

$$\therefore \quad \omega_1 AC \cos \phi = \omega_2 BC \cos \theta$$

$$\omega_1 AC \frac{AE}{AC} = \omega_2 BC \frac{BD}{BC} \qquad \left[\because \cos\phi = \frac{AE}{AC};\ \cos\theta = \frac{BD}{BC}\right]$$

$$\therefore \quad \frac{\omega_1}{\omega_2} = \frac{BD}{AE} = \frac{BP}{AP}$$

$$\left[\Delta s\ BDP \text{ and } AEP \text{ are similar } \therefore \frac{BD}{AE} = \frac{BP}{AP}\right]$$

$$\therefore \quad \frac{\omega_1}{\omega_2} = \frac{BP}{AP}$$

In Δs BDP and AEP,

$$\frac{BD}{AP} = \frac{DP}{PE}$$

$$\therefore \quad \frac{\omega_1}{\omega_2} = \frac{DP}{PE}$$

$$\omega_1 \times PE = \omega_2\, DP$$

9.4.1 Velocity of sliding

Velocity of sliding of two surfaces is the relative velocity between surfaces along common tangent $t - t$.

Velocity of sliding,

$$V_S = V_1 \sin \phi - (-V_2 \sin \theta)$$
$$= V_1 \sin \phi + V_2 \sin \theta$$

Now $V_1 = \omega_1 \times AC$

$V_2 = \omega_2 \times BC$

$$\therefore \quad V_S = (\omega_1 \times AC) \times \frac{EC}{AC} + (\omega_2 \times BC) \times \frac{DC}{BC}$$

$$\left[\because \sin\phi = \frac{EC}{AC} \text{ and } \sin\theta = \frac{DC}{BC}\right]$$

$$\therefore \quad V_S = \omega_1 \times EC + \omega_2 \times DC$$
$$\because \quad EC = PC - PE$$
$$DC = DP + PC$$
$$\therefore \quad V_S = \omega_1 \times PC - \omega_1 \times PE + \omega_2 \times DP + \omega_2 \times PC$$
$$= \omega_1 PC + \omega_2 PC \qquad [\because \omega_1 \times PE = \omega_2 \times DP]$$
$$\therefore \quad V_S = (\omega_1 + \omega_2)\, PC.$$

9.5. FORMS OF TEETH

The teeth shape should be designed to follow law of gearing. The meshing teeth are called **conjugate teeth**.

Conjugate teeth can very successfully transmit motion but these are very difficult and costly to manufacture conjugate teeth, Therefore conjugate teeth are not used. The following are the common forms of teeth used widely.

1. Cycloidal tooth profile
2. Involute tooth profile

9.5.1. Cycloidal Tooth Profile

A cycloidal is the locus of a point on the circumference of a circle which rolls without slipping on a fixed line.

Two circles 1 and 2 at point meet at C. Circle 2 is moving to the right on the pitch line without slipping. The point C traces a curve CA on the circlen. CA represents the face of the cycloidal tooth profile. When circle 1 roles to the left on the pitch line without slipping point C traces a curve CB which forms the flank of the tooth profile.

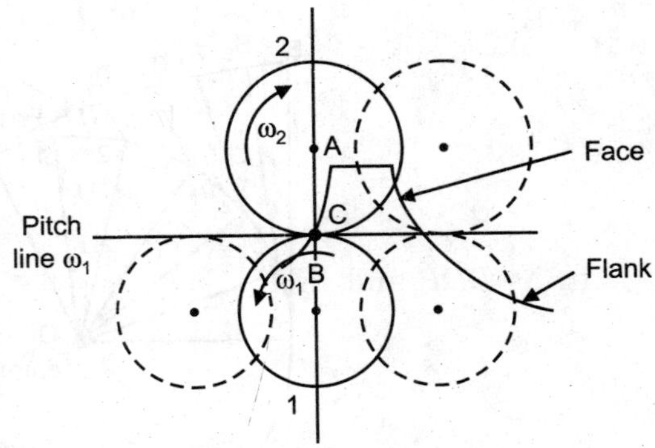

Fig. 9.4. Cycloidal Profile

Cycloidal profile may be classified as follows.

1. Epicycloid
2. Hypocycloid

1. **Epicycloid:** If a circle rolls on the external circumference of a fixed circle without slipping, a point on the circle forms a curve which is called epicyloid.

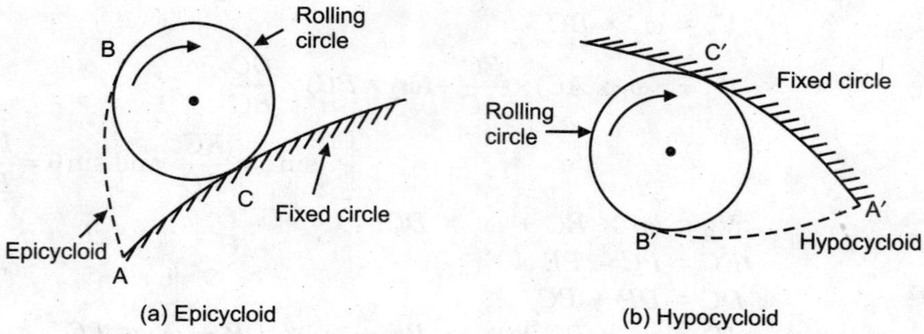

Fig. 9.5. Types of cycloids

2. **Hypocycloid:** If a circle rolls inside the circumference of a fixed circle without slipping, a point on the rolling circle traces a curve called hypocycloid.

The trace *AB* is called epicycloid and forms the face of the cycloidal tooth profile. The trace *A′B′* is called hypocycloid and forms the flank of the cycloidal tooth profile.

9.5.2. Involute Tooth Profile

Involute is the locus of a point located at the end of a thread which is unwound from the outside of a circle.

A base circle is divided into equal number of parts *A*–5, 5–4, 3–2, 2–1, 1–*B* Tangent at these points are drawn and the legnth of tangents are cut off equal in langth to arcs *B*–5, *B*–4, *B*–3, *B*–2, *B*–1. The curve $BB_1\ B_2\ B_3\ B_4\ B_5\ B'$ is the required involute.

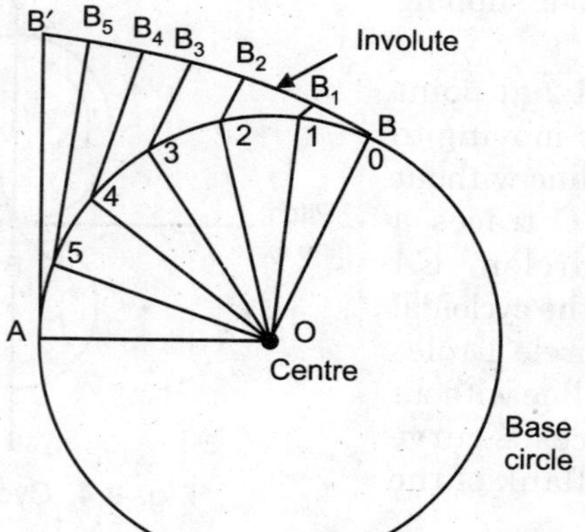

Fig. 9.6. Involute profile

9.5.3. Comparison of Cycloidal and Involute Tooth Profiles

The comparison between the two types of tooth profile is given in Table 9.1.

Table 9.1 Comparison of Gear Tooth Profiles

Sl. No.	*Cycloidal Tooth Profile*	*Involute Tooth Profile*
1.	There is smoothless smooth running of gears due to variable pressure angle.	There is smooth running of gears as the pressure angle remains constant throughout the engagement.
2.	The manufacturing of gear is complicated as the tooth has double curve, i.e., epicycloidal and hypocycloidal.	There is single curve for the teeth resulting is simplicity in the manufacturing tools.
3.	The manufacturing cost is high.	The manufacturing cost is relatively low.
4.	The centre distance must be exact to transmit motion at constant velocity ratio.	The velocity ratio is not sensitive to slight variation in centre distance
5.	There is no interference at all.	If minimum number of teeth of gear is not ensured, there can be interference of teeth.
6.	The teeth are very strong due to spreading flanks	The teeth are weaker in comparism as teeth have radial flanks.
7.	There is less wear due to mating of a convex flank and concave face.	There are chances of wear due to mating of convex surfaces of flanks and faces.

9.6. SYSTEMS OF GEAR TEETH

The tooth system is a standard to specify relationship between addendum, dedendum, clearance, tooth thickness, and fillet radius to attain interchangeability of gears of all tooth numbers but of the same pressure angle of diametral pitch and module. Standards are defined by the American Gear Manufactures Association (AGMA) and the American National Standards Institute (ANSI).

9.6.1. Standard Tooth Sizes

The standard tooth sizes are given in Table 9.2.

Table 9.2 Standard Tooth Sizes

	Standard modules (mm/tooth)
Preferred	1,1.25, 1.5, 2, 2.5, 3, 4, 5, 6, 7, 8, 10, 12, 16, 20, 25, 32, 40, 50
Next choice,	1, 1.25, 1.375, 1.75, 2.25, 2.75, 3.5, 4.5, 5.5, 7, 9, 11, 14, 18, 22, 28, 36, 45

9.6.2. Standard Tooth Systems

Four standard tooth systems are used as given in Table 9.3.

Table 9.3. Standard Tooth Systems for Spur Gears

System	*Pressure angle ϕ (deg)*	*Addendum, a*	*Dedndum, d*
Full depth	20°	1 m	1.25 m
Full depth	22.5°	1 m	1.25 m
Full depth	25°	1 m	1.25 m
Sub teeth	20°	0.8 m	1 m

9.7. LENGTH OF PATH OF CONTACT

The length of path of contact is the distance between the point where the contact of two teeth (pinion and wheel) begins to the point where the contact ends. The path of contact is the sum of path of approach and path of recess.

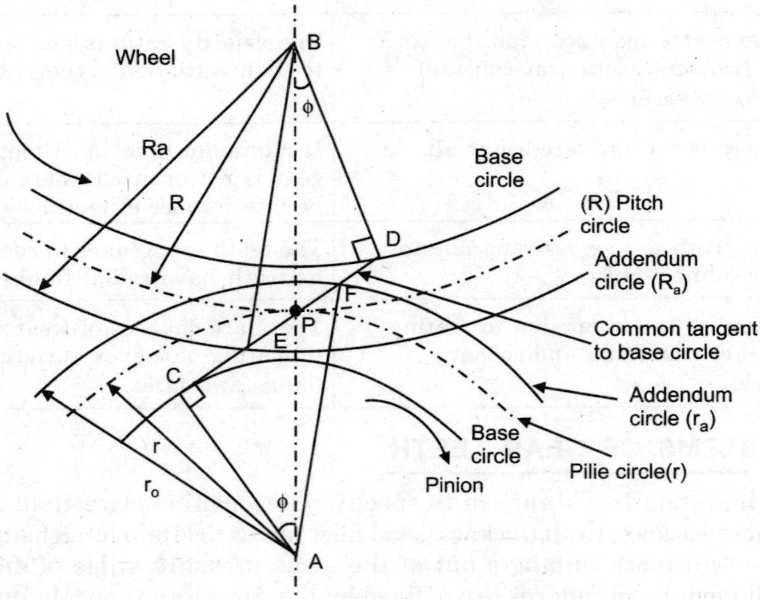

Fig. 9.7. Gear Teeth Action

A pinion with centre A and rotating clockwise drives a wheel with centre B in anticlockwise direction.

The following are salient features of Fig. 9.7.

P = Point of meeting of pitch circle

CD = Common tangent to basic circle
= Line of action

E = Point of intersection of common tangent and addendum circle of wheel.

F = Point of intersection of common tangent and addendum circle of pinion.

EF = Length of path of contact
EP = Path of approach.
PF = Path of recess

Assume, R = Pitch circle radius of wheel = BP
r = Pitch circle radius of pinion = PA
R_a = Addendum circle radius of wheel = BE
r_a = Addendum circle radium of pinion = AF

$$\therefore \quad EF = EP + PF$$
$$= (ED - PD) + (CF - CP)$$

In right angle ΔBPD,

$$PD = BP \sin \phi = R \sin \phi$$
$$BD = BP \cos \phi = R \cos \phi$$

In right angle ΔACP,

$$CP = \text{PA} \sin \phi = r \sin \phi$$
$$CA = \text{PA} \cos \phi = r \cos \phi$$

From right angle ΔBED,

$$ED = \sqrt{(BE)^2 - (BD)^2} = \sqrt{R_a^2 - R^2 \cos^2 \phi}$$

From right angle ΔFCA

$$CF = \sqrt{(AF)^2 - (CA)^2} = \sqrt{r_a^2 - r^2 \cos^2 \phi}$$

$\therefore$ Path of approach,

$$EP = ED - PD$$
$$= \left(\sqrt{R_a^2 - R^2 \cos^2 \phi}\right) - R \sin \phi$$

Path of recess

$$PF = CF - CP$$
$$= \left(\sqrt{r_a^2 - r^2 \cos^2 \phi}\right) - r \sin \phi$$

Hence
$$EF = \left\{\left(\sqrt{R_a^2 - R^2 \cos^2 \phi}\right) - R \sin \phi\right\} + \left\{\left(\sqrt{r_a^2 - r^2 \cos^2 \phi}\right) - r \sin \phi\right\}$$
$$= \sqrt{R_a^2 - R^2 \cos^2 \phi} + \sqrt{r_a^2 - r^2 \cos^2 \phi} - (R - r) \sin \phi$$

Note: The path of approach can be found out from dimensions of wheel and path of recess can be found out from dimensions of pinion.

9.8. CONTACT RATIO

Contact ratio is the number of pairs of teeth in contact. This is the ratio of length of arc of contact to the circular pitch.

$$\text{Contact ratio} = \frac{\text{Length of arc of contact}}{\text{Cicular pitch}}$$

$$\text{Length of arc of contact} = \frac{\text{Length of path of contact}}{\cos \phi} = \frac{EF}{\cos \phi}$$

Circular pitch, $p_c = \pi m$

where m = madule

$\therefore$ Contact ratio $= \dfrac{EF}{\cos\phi} \times \dfrac{1}{\pi m} > 1$.

Example 9.2. **A gear wheel has 60 teeth pitch circle dimeter of 750 mm. Find:**

(*a*) circular pitch

(*b*) module of teeth

Solution : (*a*) Circular pitch, $p_c = \dfrac{\pi d}{T} = \dfrac{\pi \times 750}{60}$ **= 393 mm Ans.**

(*b*) Madule, $m = \dfrac{d}{T} = \dfrac{750}{60}$ **= 12.5 mm Ans.**

Example 9.3. **Two wheels have 18 and 45 teeth respectively. If the circular pitch is 50 mm, find pitch circle radii.**

Solution : Pitch circle, $p_c = \dfrac{2\pi r}{T}$

$$r = \frac{p_c T}{2\pi}$$

$\therefore$ $r_1 = \dfrac{p_c T_1}{2\pi} = \dfrac{50 \times 18}{2\pi}$ **= 143.2 mm Ans.**

$r_2 = \dfrac{p_c T_2}{2\pi} = \dfrac{50 \times 45}{2\pi}$ **= 358 mm Ans.**

Example 9.4. **A gear wheel with 96 teeth has module of 1.25. Determine.**

(*i*) Pitch circle diameter

(*ii*) Diametral pitch

(*iii*) Circular pitch

Solution : (*i*) Module, $m = \dfrac{d}{T}$

$\therefore$ $d = mT = 1.25 \times 96 =$ **120 mm Ans.**

(*ii*) Diametral pitch,

$p_d = \dfrac{T}{d} = \dfrac{96}{120}$ **= 0.8 tooth/mm Ans.**

(*iii*) Circular petch, $p_c = \dfrac{\pi}{p_d} = \dfrac{\pi}{0.8}$ **= 3.93 mm Ans.**

Example 9.5. **A spur gear of 30 teeth rotates at 200 rpm. It has a module of 2 mm. Find:**

(*a*) Circular patch

(*b*) Pitch line velocity.

Solution : (*a*) Cicular pitch,

$$p_c = \frac{\pi d}{T} = \pi\text{m}$$

$$= \pi \times 2$$

$=$ **6.28 mm Ans.**

(*b*) Pitch line velocity,

$$v_p = \omega r$$

$$= \frac{2\pi N}{60} \times \frac{d}{2} = \frac{2\pi \times 200}{60} \times \frac{mT}{2}$$

$$= \frac{2\pi \times 200}{60} \times \frac{2 \times 30}{2}$$

$$= \mathbf{628.3\ mm/s\ Ans.}$$

Example 9.6. Two meshing gears have the following data:

Velocity ratio, $VR = \frac{1}{3}$
Module, m = 4 mm
Pressure angle, ϕ = 20°
Centre distance, C = 200 mm

Find:
(*a*) Number of teeth
(*b*) Base circle radius of gear wheel.

Solution :

(*a*) Velocity ratio, $VR = \frac{N_2}{N_1} = \frac{T_1}{T_2} = \frac{1}{3}$

$\therefore \quad T_2 = 3T_1$

Centre distance, $C = \frac{d_1 + d_2}{2} = \frac{m(T_1 + T_2)}{2}$

$$= \frac{m(T_1 + 3T_1)}{2} = \frac{m \times 4T_1}{2}$$

$\therefore \quad T_1 = \frac{2C}{4m} = \frac{2 \times 200}{4 \times 4} = 25$

$T_2 = 3 \times 25 = \mathbf{75}$

(*b*) Pitch circle diameter, $d_2 = mT_2 = 4 \times 75 = 300$ mm

Base circle radius $= \frac{d_2}{2}\cos\phi = \frac{300}{2}\cos 20° =$ **141 mm Ans.**

Example 9.7. Two wheels with 28 and 45 teeth have standard addendem of one module. Find the length of path of contact and length of arc of contact if the pressure angle is 20°.

Solution : Assume $T_1 = 28$

$T_2 = 45$

Addendum, $a = 1$ m

Pressure angle $\phi = 20°$

Radius of pitch circle of pinion, $r = \frac{mT_1}{2} = \frac{p_c T_2}{2\pi}$

$$= \frac{28 p_c}{2\pi} = 4.48\ p_c$$

Radius of pitch circle of wheel $R = \frac{p_c T_2}{2\pi} = \frac{45 \times p_c}{2\pi} = 7.16\ p_c$

Addendum radius of pinion,

$$r_a = r + a = r + m$$

$$= 4.48\ p_c + m = 4.48\ p_c + \frac{p_c}{\pi}$$

$$= 4.79\ p_c$$

Addendum radius of gear wheel, $R_a = R + a$

$$R_a = 7.16\ p_c + \frac{p}{\pi} = 7.47\ p_c.$$

(*a*) Lenght of path of contact

$$= \sqrt{R_a^2 - R^2 \cos^2 \phi} + \sqrt{r_a^2 - r^2 \cos^2 \phi} - (R - r)\cos\phi$$

$$= \sqrt{(7.47\ p_c)^2 - (7.16 p_c)^2 \cos^2 \phi} + \sqrt{(4.79 p_c)^2 - (4.48 p_c)^2 \cos^2 \phi} - (7.16 p_c - 4.48\ p_c) \cos \phi$$

$$= \mathbf{1.59\ p_c}\ \textbf{Ans.}$$

(*b*) Length of arc of contact

$$= \frac{\text{Length of path of contact}}{\cos\phi}$$

$$= \frac{1.59 p_c}{\cos 20°} = \mathbf{1.69\ p_c}\ \textbf{Ans.}$$

9.9. INTERFERENCE AND UNDERCUTTING IN INVOLUTE GEAR TEETH

Interference is the problem faced by non-conjugate gears when the lip of one tooth undercuts the root of the other mating teeth. Interference can lead to following problems.

1. There is no proper sliding of the non-conjugate non-involute teeth and hence no proper mating.
2. The contacting teeth may have different velocities leading to locking of two mating gears.
3. There can be binding of gears and rough action of motion.

Undercutting is the removal of material due to interference. This can be avoided by proper design of the teeth to avoid interference. Undercutting can result in weakening of the tooth which can be very serious in case of teeth of pinion where number of teeth are small.

The action of two mating gears is shown in Fig. 9.8.

CD = path of contact between the teeth

The teeth or the pinion and wheel are engazed at D and disengaged at C.

Case 1. The radius of addendum circle of the pinion is increased so that point C shifts to point M. The tip of the tooth on the pinion will undercut the tooth on the wheel at the root and remove material of tooth of the wheel. Hence there is interference in mating.

Case 2. The radius of the addendum circle of the wheel is increased. The lip of the tooth will cause interference with the tooth of the pinion.

Hence interference can be avoided if the addendum circle of the wheel and pinion intersect the line of action between M and N.

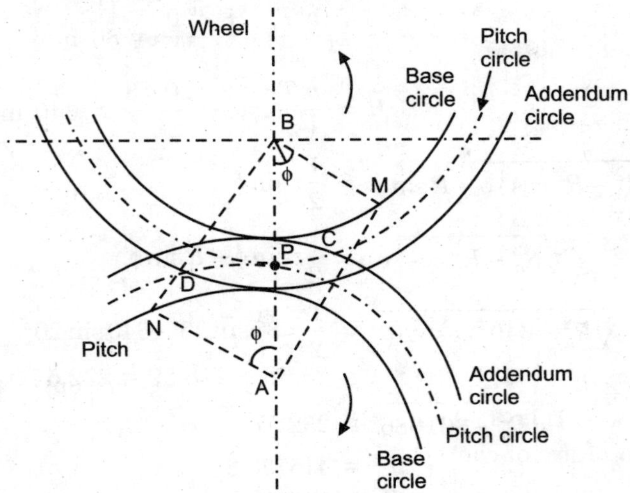

Fig. 9.8. Interference and undercutting.

Pitch circle radius of pinion, $r = AP$.
Pitch circle radius of wheel, $R = BP$.
In right angle ΔAPN,

$$PN = AP \sin\phi = r \sin\phi$$
$$= \text{Max. length of path of approach.}$$

In right angle ΔBMP

$$PM = BP \sin\phi = R \sin\phi$$
$$= \text{Max. length of recess}$$

Max. length of path of contact,

$$MN = PN + PM = r \sin\phi + R \sin\phi$$
$$= (r + R) \sin\phi$$

$$\text{Max. length of arc of contact} = \frac{(r+R)\sin\phi}{\cos\phi} = (r + R) \tan\phi.$$

Example 9.8. A pinion of 20° involute teeth rotating at 275 rpm meshes with a gear and provides a gear ratio of 1:8. The number of teeth on the pinion is 20 and the module is 8 mm. If the interference is just avoided, determining

(*i*) The addenda on the wheel and pinion
(*ii*) The path of contact and
(*iii*) The maximum velocity of sliding on both sides of the pitch point.
[UPTU 2010-11]

Solution :

$$\phi = 20°$$
$$GR = 8$$
$$T_1 = 20$$
$$N_1 = 275 \text{ rpm}$$

$$m = 8 \text{ mm}$$

(i) Addendum for pinion and gear wheel

$$r = \frac{mT_1}{2} = \frac{8 \times 20}{2} = 80 \text{ mm}$$

$$R = \frac{mT_2}{2} = \frac{8 \times 20 \times 8}{2} = 640 \text{ mm}$$

$$\sqrt{R_a^2 - R^2\cos^2\phi} - R\sin\phi = \frac{1}{2} r\sin\phi$$

$$\sqrt{R_a^2 - R^2\cos^2\phi} = \frac{1}{2} r\sin\phi + R\sin\phi$$

$$\sqrt{R_a^2 - 640^2\cos^2 20°} = \frac{1}{2} \times 80\sin 20° + 640\sin 20°$$

$$= 13.68 + 218.89 = 232.57$$

$$\sqrt{R_a^2 - 361686} = 232.57$$

$$R_a^2 = 415774.8$$

$$R_a = 644.8 \text{ mm}$$

$$\text{Addendum of wheel} = R_a - R = 644.8 - 640 = 4.8 \text{ mm.}$$

Now $\sqrt{r_a^2 - r^2\cos^2\phi} - r\sin\phi = \frac{1}{2} R\sin\phi$

$$\sqrt{r_a^2 - 80^2\cos^2 20°} = \frac{1}{2} \times 640\sin 20° + 80\sin 20°$$

$$\sqrt{r_a^2 - 5651.34} = 109.446 + 27.36 = 136.80$$

$$r_a^2 = 18716.32 + 5651.34 = 24367.66$$

$$r_a = 156 \text{ mm.}$$

$$\text{Addendum of pinoin} = 156 - 80 = 76 \text{ mm}$$

(ii) Path of contact

$$\text{Path of contact} = \frac{1}{2} \text{ Max. length of path of approach} + \frac{1}{2} \text{ Max. length of path of recess}$$

$$= \frac{1}{2} r\sin\phi + \frac{1}{2} R\sin\phi$$

$$= \frac{1}{2}(r + R)\sin\phi$$

$$= \frac{1}{2}[80 + 640]\sin 20° = \mathbf{123.13 \text{ mm}} \textbf{ Ans.}$$

(iii) Max. velocity of sliding during approach and recess

$$\omega_1 = \frac{2\pi N_1}{60} = \frac{2\pi \times 275}{60} = 28.8 \text{ rad/s}$$

$$\omega_2 = \omega_1 \times \frac{T_1}{T_2} = 28.8 \times \frac{20}{160} = 3.6 \text{ rad/s}$$

Max. velocity of sliding during approach

$$= (\omega_1 + \omega_2) \times \text{Length of path of approach}$$

$$= (28.8 + 3.6) \times \frac{r \sin \phi}{2}$$

$$= 32.4 \times \frac{80 \sin 20°}{2} = \mathbf{44.325\ mm/s\ Ans.}$$

Max, velocity of sliding during recess

$$= (\omega_1 + \omega_2) \times \text{Length of recess}$$

$$= 32.4 \times \frac{1}{2}\ R \sin \phi$$

$$= 32.4 \times \frac{1}{2} \times 640 \sin 20°$$

$$= \mathbf{3546\ mm/s\ Ans.}$$

9.10. MINIMUM NUMBER OF TEETH ON PINION TO AVOID INTERFERENCE

The minimum number of teeth on pinion can be found out by satisfying the limiting condition to avoid interference.

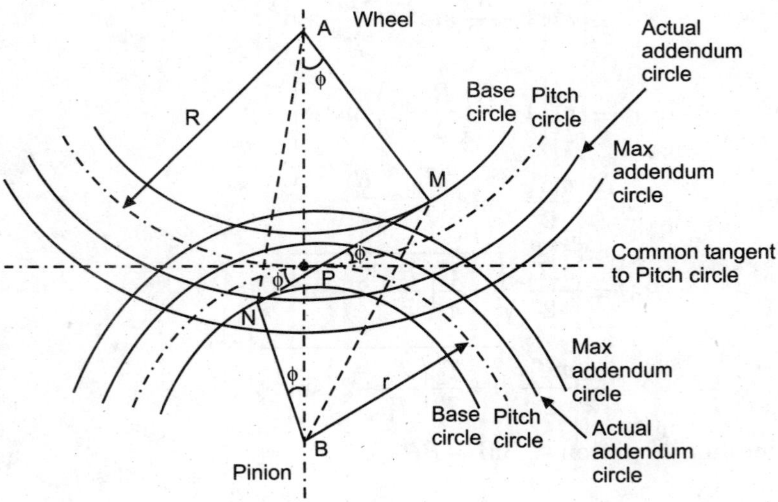

Fig. 9.9. Minimum Pinion Teeth

Assume T_2 = No. of teeth on wheel

T_1 = No. of teeth on pinion

m = Module of teeth

d = Pitch circle diameter of pinion

$= 2r = m \times T_1$

r = Pitch circle radius of pinion

$$= \frac{d}{2} = \frac{m \times T_1}{2}$$

R = Pitch circle radius of wheel

$$= \frac{D}{2} = \frac{m \times T_2}{2}$$

$$G = \text{Gear ratio} = \frac{T_2}{T_1} = \frac{R}{r}$$

ϕ = Pressure angle = Angle of obliquity

A = Cente of wheel

B = Centre of pinion

From ΔMPB,

$$BM^2 = BP^2 + PM^2 - 2BP \times PM \times \cos BPM$$

$$= r^2 + R^2 \sin^2 \phi - 2r \times R \sin \phi \times \cos (90° + \phi)$$

$\because$ $BP = r$

$PM = R \sin \phi$

$\angle BPM = 90° + \phi$

Now $\cos (90° + \phi) = - \sin \phi$

$$\therefore \quad BM^2 = r^2 + R^2 \sin^2 \phi - 2Rr \sin \phi \times (- \sin \phi)$$

$$= r^2 + R^2 \sin \phi + 2Rr \sin^2 \phi$$

$$= r^2 \left[1 + \frac{R^2}{r^2} \sin^2 \phi + \frac{2R}{r^2} \sin^2 \phi\right]$$

$$= r^2 \left[1 + \frac{R}{r}\left(\frac{R}{2} + 2\right) \sin^2 \phi\right]$$

Now $r = \dfrac{mT_1}{2}$ and $\dfrac{T_2}{T_1} = \dfrac{R}{r} = G$

$$\therefore \quad BM = \frac{mT_1}{2} \sqrt{1 + \frac{R}{r}\left(\frac{R}{r} + 2\right) \sin^2 \phi}$$

$$= \frac{mT_1}{2} \sqrt{1 + \frac{T_2}{T_1}\left(\frac{T_2}{T_1} + 2\right) \sin^2 \phi} \qquad \text{...(1)}$$

Addendum of pinion = $BM - BP$

$$= BM - r = BM - \frac{m \times T_1}{2}$$

$$= \frac{mT_1}{2} \sqrt{1 + \frac{T_2}{T_1}\left(\frac{T_2}{T_1} + 2\right) \sin^2 \phi} - \left(\frac{mT_1}{2}\right) \qquad \text{...(2)}$$

From equations (1) and (2)

$$\text{Addendum of pinion} = \left\{\frac{m \times T_1}{2} \sqrt{1 + \frac{T_2}{T_1}\left(\frac{T_2}{T_1} + 2\right) \sin^2 \phi - 1}\right\}$$

Let $A_p \times m$ = Addendum of pinion to avoid interference; where A_p is the multiplying factor for module to avoid interference.

Addendum of pinion = $A_p \times m$

$$A_p \times m = \frac{mT_1}{2}\left[\sqrt{1+\frac{T_2}{T_1}\left(\frac{T_2}{T_1}+2\right)\sin^2\phi}-1\right]$$

$$A_p = \frac{T_1}{2}\left[\sqrt{1+\frac{T_2}{T_1}\left(\frac{T_2}{T_1}+2\right)\sin^2\phi}-1\right]$$

or
$$T_1 = \frac{2A_p}{\left[\sqrt{1+G(G+2)\sin^2\phi}-1\right]}$$

where T_1 is the number of minimum number of teeth on pinion to avoid interference.

Note. 1. For $G = 1$ or $T_1 = T_2$

$$T_1 = \frac{2A_p}{\left[\sqrt{1+1(1+2)\sin^2\phi}-1\right]} = \frac{2A_p}{\left[\sqrt{1+3\sin^2\phi}-1\right]}$$

2. The minimum number of teeth on pinion which will mesh any gear or rack without interferenc are given in Table 9.3.

Table 9.4: Minimum Number of Teeth on Pinion

Sl. No.	*System of gear teeth*	*Min. no. of pinion teeth*
1.	$14\frac{1}{2}^\circ$ composite	12
2.	$14\frac{1}{2}^\circ$ full depth involute	32
3.	20° full depth involute	18
4.	20° stub involute	14

9.11. MINIMUM NUMBER OF TEETH ON WHEEL TO AVOID INTERFERENCE

Refer to Fig. 9.10,

$$\begin{aligned} AN^2 &= AP^2 + PN^2 - 2AP \times PN \cos APN \\ &= R^2 + r^2\sin^2\phi - 2Rr\sin\phi\cos(90^\circ+\phi) \\ &= R^2 + r^2\sin^2\phi + 2Rr\sin\phi \\ &= R^2\left[1+\frac{r^2}{R^2}\sin^2\phi+\frac{2Rr\sin^2\phi}{R}\right] \\ &= R^2\left[1+\frac{r}{R}\left(\frac{r}{R}+2\right)\sin^2\phi\right] \end{aligned}$$

$$\therefore \qquad AN = R\sqrt{\left[1+\frac{r}{R}\left(\frac{r}{R}+2\right)\sin^2\phi\right]}$$

Addendum of the wheel, $= AN - AP$

$$= AN - R$$

$$= AN - \frac{mT_2}{2}$$

$$= R\sqrt{1+\frac{r}{R}\left(\frac{r}{R}+2\right)\sin^2\phi} - \frac{mT_2}{2}$$

$$= \frac{mT_2}{2}\sqrt{1+\frac{r}{R}\left(\frac{r}{R}+2\right)\sin^2\phi} - \frac{mT_2}{2}$$

$$\therefore \quad \text{Addendum on wheel} = \frac{mT_2}{2}\left[\sqrt{1+\frac{r}{R}\left(\frac{r}{R}+2\right)\sin^2\phi} - 1\right]$$

$$= \frac{mT_2}{2}\left[\sqrt{1+\frac{1}{G}\left(\frac{1}{G}+2\right)\sin^2\phi} - 1\right]$$

Let $A_W \times m$ = Addendum of wheel to avoid interference.
Addendum of wheel = $A_W \times m$

$$\therefore \qquad A_W \times m = \frac{m \times T_2}{2}\sqrt{1+\frac{r}{R}\left(\frac{r}{R}+2\right)\sin^2\phi} - \frac{mT_2}{2}$$

$$\therefore \qquad A_W = \frac{T_2}{2}\left[\sqrt{1+\frac{r}{R}\left(\frac{r}{R}+2\right)\sin^2\phi} - 1\right]$$

$$\therefore \qquad T_2 = \frac{2 \times A_W}{\sqrt{1+\frac{r}{R}\left(\frac{r}{R}+2\right)\sin^2\phi} - 1}$$

$$= \frac{2 \times A_W}{\sqrt{1+\frac{1}{G}\left(\frac{1}{G}+2\right)\sin^2\phi} - 1}$$

where T_2 = Min. number of teeth on wheel to avoid interference.

Note 1. For $G = 1$

$$T_2 = \frac{2A_W}{\sqrt{1+3\sin^2\phi} - 1}$$

or

$$T_2 = \frac{2 \times A_W}{G\left[\sqrt{1+\frac{1}{G}\left(\frac{1}{G}+2\right)\sin^2\phi} - 1\right]}.$$

Example 9.8. **A wheel and pinion with 50 and 13 involute teeth of module 10 mm and 20° pressure angle have addendum of one module. Find.**

(*a*) Does interference occur?

(*b*) What should be the pressure angle to avoid interference.

Solution :

(*a*) Pitch radius of wheel,

$$R = \frac{mT_2}{2} = \frac{10 \times 50}{2} = 250 \text{ mm}$$

Pitch radius of pinion

$$r = \frac{mT_1}{2} = \frac{10 \times 13}{2} = 65 \text{ mm}$$

Addendum radius of wheel,

$$R_a = R + \text{Addendum} = 250 + 10 = 260 \text{ mm}$$

$$\text{Actual path of approach} = \sqrt{R_a^2 - R^2 \cos^2 \phi} - R \sin \phi$$

$$= \sqrt{260^2 - 250^2 \cos^2 20°} - 250 \sin 20°$$

$$= 25.91 \text{ mm}$$

$$\text{Max. length of path of approach} = r \sin \phi$$

$$= 65 \sin 20° = 22.23 \text{ mm}$$

Interference will occur as actual path of approach (25.91 mm) is more than max. length of path of approach (22.23 mm)

(*b*) Let max permissible length = actual length of path of approach

$$r \sin \phi = 25.91$$

$$65 \sin \phi = 25.91$$

$$\sin \phi = \frac{25.91}{65} = 0.3986$$

$$\therefore \quad \phi = 23.49° \text{ Ans.}$$

Example 9.9. **A pair of spur gears have the following data:**

Gear ratio, $G = 4 : 1$

Pressure angle $\phi = 14°$

Arc of approach $\nless$ circular pitch

Find (*a*) Minimum number of teeth on pinion and wheel.

(*b*) Addendum of wheel.

Solution : $G = \dfrac{T_2}{T_1} = \dfrac{R}{r} = 4$

(*a*) Least number of teeth on each wheel

$$\text{Max. length of arc of approach} = \frac{\text{Max. length of path of approach}}{\cos \phi}$$

$$= \frac{r \sin \phi}{\cos \phi} = r \tan \phi.$$

$$p_c = \pi \times m = \pi \times \frac{2r}{T_1}$$

$$r \tan \phi = p_c = \pi \times \frac{2r}{T_1}$$

$$\therefore \quad T_1 = \frac{\pi \times 2r}{r \tan \phi} = \frac{2\pi}{\tan \phi} = \frac{2\pi}{\tan 14.5°} = 24.3 \approx 25.$$

$$T_2 = G \times T_1 = 4 \times 25 = \mathbf{100} \quad \textbf{Ans..}$$

(*b*) Addendum of wheel

$$\text{Addendum of wheel} = \frac{m \times T_2}{2}\left[\sqrt{1 + \frac{1}{G}\left(\frac{1}{G} + 2\right)\sin^2 \phi} - 1\right]$$

$$= \frac{m \times 100}{2}\left[\sqrt{1 + \frac{1}{4}\left(\frac{1}{4} + 2\right)\sin^2 14.5°} - 1\right]$$

$$= 0.8735 \text{ m} = 0.8735 \frac{p_c}{\pi} = \mathbf{0.278\ p_c} \textbf{ Ans.}$$

QUESTION BANK

1. Give a detailed classification of gears.
2. Define the following terms
   (*i*) Pitch circle
   (*ii*) Pitch diameter
   (*iii*) Pitch point
   (*iv*) Circular pitch
   (*v*) Module
3. State and derive the law of gearing. *[UPTU 2010-11]*
4. Dirive an expression for velocity of sliding.
5. Make a comparison of cycloidal and involute teeth profiles.
6. How is interchangeability of gears ensured?
7. What is path of contact ? Deduce an expression to find its value.
8. How to find minimum no. of teeth to avoid interference?
9. Explain the term interference and undercutting of gear teeth.

NUMERICAL PROBLEMS

1. Find the least number of teeth if pressure angle is 20° and arc of approach of recess is not less than the circular pitch. **[Ans.** 18]
2. Two spur gears have a velocity ratio of 2.5 and pressure angle of 20°, module of involute teeth is 4 mm and addendum equal to 1.23 module, pinion rotates at 150 rpm. Find:
   (*a*) Minimum number of teeth on each wheel to avoid interference.
   (*b*) Number of pairs of teeth in contact. **[Ans.** (*a*) 45, 18 (*b*) 1.95]
3. A spur gear has 30 teeth, moduls of 1.4 mm and speed of rotation 360 rpm. Find
   (*a*) Circular pitch
   (*b*) Pitch line velocity **[Ans.** (*a*) 4.4 mm (*b*) 791.7 mm/s]

❑❑❑

Chapter 10

GEAR TRAINS

10.1. INTRODUCTION

In machines and mechanism, it is very common to transmit power and motion from one shaft to another. Two or more gears are arranged in series or parallel combinations to achieve the above aim. Such combinations are called gear trains. There are different types of gear trains using spur, bevel or spiral gears. The nature of the train used depends upon the velocity ratio required and relative position of the axes of the shafts. Gear trains are very useful to obtain large velocity ratio within a small space. Gear trains are very widely used in machine tools, automobiles, hoists, wrist watches, etc.

10.2. TYPES OF GEAR TRAINS

There are different types of gear trains depending upon the arrangement of gear wheels.

1. *Simple gear train.*

   There is only one gear on each shaft.

2. *Compound gear train*

   There can be more than one gear on a shaft.

3. *Reverted gear train*

   The driving gear and driven gear are co-axial.

   In the above types of gear trains, the axis of rotation of the wheels are fixed in position and gears rotate about their respective axes.

4. *Planetary gear trains*

   In some gear trains, the axes of rotation of gear wheel are not fixed and may rotate around the axes of other wheel with which they mesh. Such trains are known as planetary gear trains or epicyclic gear trains. Such trains are useful to transmit very high velocity ratios using small gears accommodated in very less space.

5. *Sun and planet gear*

   A compound epicyclic gear train with an annulus wheel is called sun and planet gear. The sun gear is co-axial with the annulus gear and the motion resembles that of a planet around the sun.

10.3 SIMPLE GEAR TRAIN

A simple gear train consisting of a pair of external gears is shown in Fig. 10.1. The gears are represented by their pitch circle.

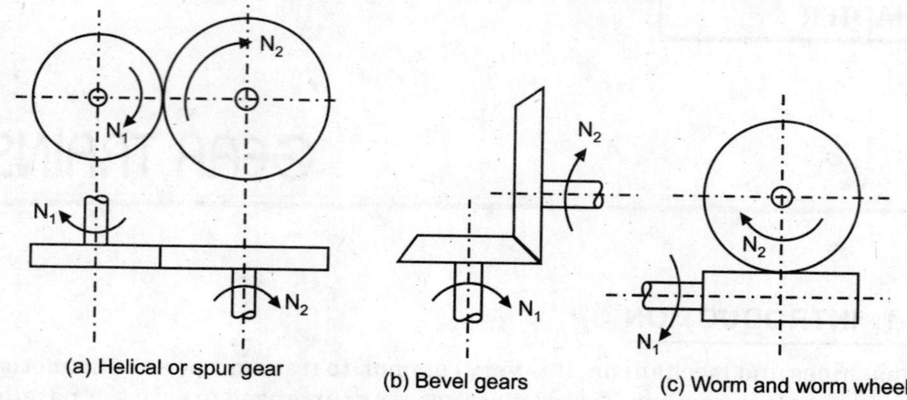

Fig. 10.1. Simple Gear Trains

Gear 1. The gear 1 drives gear 2. Therefore gear 1 is called the driver or the pinion or the worm.

Gear 2. The driven gear 2 is called the driven or the follower or simple gear or worm wheel.

N_1 = Speed of the pinion (or driver) in rpm.

N_2 = Speed of the gear (or follower) in rpm

T_1 = Number of teeth on gear 1.

T_2 = Number of teeth on gear 2.

10.3.1. Speed Ratio

The speed ratio of the gear train is the ratio of the speed of driver to the speed of follower.

$$\text{Speed ratio} = \frac{N_1}{N_2}$$

This is also called velocity ratio.

Applying the law of gearing,

$$\text{Speed ratio} = \frac{N_1}{N_2} = \frac{T_2}{T_1}$$

10.3.2. Train Value

Train value is the ratio of speed of follower to the speed of driver.

$$\text{Train value} = \frac{N_2}{N_1} = \frac{T_1}{T_2}$$

This is also called the kinematic coefficient. The train value is the reciprocal of speed ratio.

The speed ratio is negative because the direction of rotation of follower or output shaft is opposite to that of input shaft.

10.3.3. Use of Intermediate Gears

Intermediate gears are used to achieve the following:

1. To change the direction of rotation of the output shaft.
2. To connect gears where the input and output shaft are located at large distance. This helps to use smaller pinion and gear more economically.

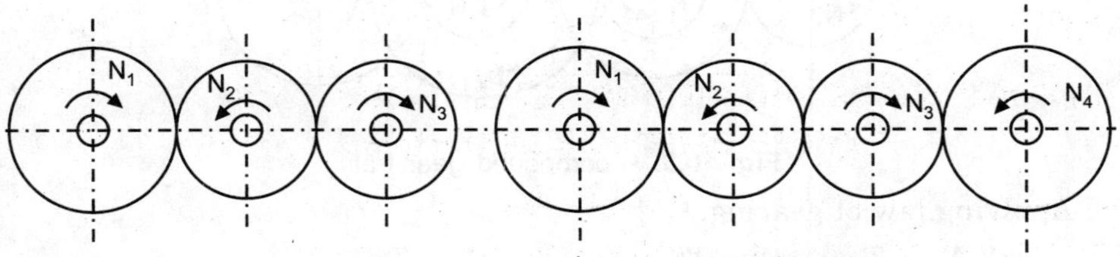

Fig. 10.2. Use of Intermediate Gears

In order to achieve same directions of rotation of input end output shafts, one intermediate shaft is used. In order to connect gears at large distances more gears may be used. If number of intermediate gears is odd, the speed ratio is called positive whereas when number of intermediate gears is even, the speed ratio is negative.

Using law of gearing,

$$\frac{N_1}{N_2} = \frac{T_2}{T_1}$$

$$\frac{N_2}{N_3} = \frac{T_3}{T_2} \quad \text{and} \quad \frac{N_4}{N_3} = \frac{T_3}{T_4} \quad ....$$

Multiplying,

$$\frac{N_1}{N_2} \times \frac{N_2}{N_3} \times \frac{N_3}{N_4} = \frac{T_2}{T_1} \times \frac{T_3}{T_2} \times \frac{T_4}{T_3}$$

$$\therefore \qquad \text{Speed ratio} = \frac{\text{Speed of input shaft (1)}}{\text{speed of output shaft (4)}}$$

$$= \frac{N_1}{N_4} = \frac{T_4}{T_1}$$

$$\text{Train value} = \frac{T_1}{T_4}$$

The speed ratio and train value are independent of number and size of intermediate gears. Therefore, intermediate gears are also called idlers.

10.4. COMPOUND GEAR TRAIN

A compound gear train contains a series of gears. It may contain only one type of gears with more than one gear on a shaft or a combination of bevel, helical, spur gears. In such trains normally some of the intermediate shafts may carry more than one gear except the input and output shafts.

A compound gear train consisting of driver 1, follower 6 and two compound gears 2 and 3 and 4 and 5 is shown in Fig. 10.3.

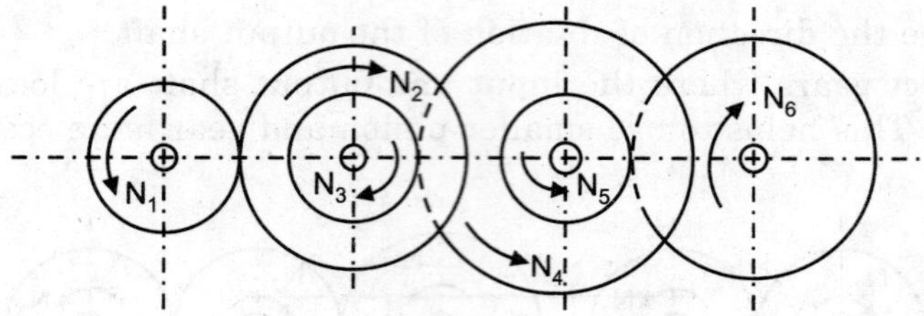

Fig. 10.3. A compound gear train

Applying law of gearing,

$$\frac{N_1}{N_2} = \frac{T_2}{T_1}; \quad \frac{N_3}{N_4} = \frac{T_4}{T_3}; \quad \text{and} \quad \frac{N_5}{N_6} = \frac{T_6}{T_5}$$

Multiplying,

$$\frac{N_1}{N_2} \times \frac{N_3}{N_4} \times \frac{N_5}{N_6} = \frac{T_2}{T_1} \times \frac{T_4}{T_3} \times \frac{T_6}{T_5}$$

Now $N_2 = N_3$ and $N_4 = N_5$

$$\therefore \quad \frac{N_1}{N_2} \times \frac{N_2}{N_4} \times \frac{N_4}{N_6} = \frac{T_2}{T_1} \times \frac{T_4}{T_3} \times \frac{T_6}{T_5}$$

or Speed ratio, $= \dfrac{N_1}{N_6} = \dfrac{T_2}{T_1} \times \dfrac{T_4}{N_3} \times \dfrac{T_6}{T_5}$

or Train value $= \dfrac{N_6}{N_1} = \dfrac{T_1}{T_2} \times \dfrac{T_3}{T_4} \times \dfrac{T_5}{T_6}$

$$= \frac{\text{Product of number of teeth on driving gears}}{\text{Product of number of teeth on driven gears}}$$

$$= \frac{\text{Speed of the last follower}}{\text{Speed of the first driver}}$$

Note: 1. The meshing gears must have the same module or circular pitch. In the above example, gears 1 and 2 much have same module. Similarly gear 3 and 4 and 5 and 6 must have same module.

2. In compound gear train, very large speed ratios can be achieved with the help of small gears as compared to simple gear trains.
3. For a speed ratio above 7, a simple gear train is not economical. For such speed ratio a compound gear train or worm gearing is employed.

10.5. REVERTED GEAR TRAIN

In a reverted gear train, the first and last gears have collinear axes of rotation. These arrangements are used in speed reducers for compactness, in clocks to connect hour hand to the minute hand and in simple lathes where back gear is used. A reverted gear train is shown in Fig. 10.4.

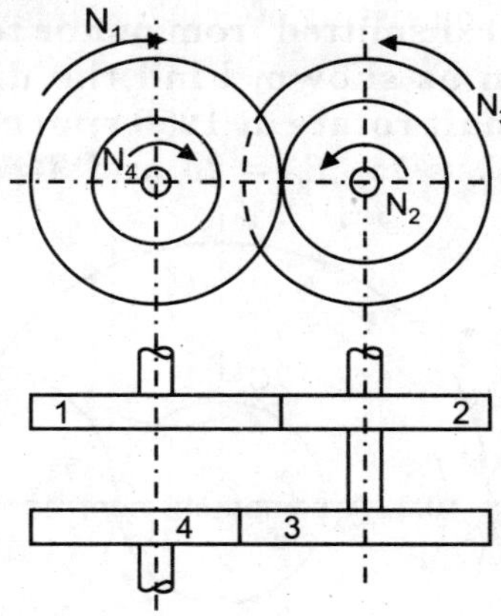

Fig. 10.4. A Reverted Gear Train

In a reverted gear train, the speed ratio is positive

T_1, T_2, T_3, T_4 = Number of teeth on respective gears

N_1, N_2, N_3, N_4 = Speed of respective gears

r_1, r_2, r_3, r_4 = Pitch circle radii of respective gears

For same distance between centres of gears 1 and 2 and 3 and 4,

$$r_1 + r_2 = r_3 + r_4$$

$$\therefore \quad T_1 + T_2 = T_3 + T_4$$

$\because$ Number of teeth on each gear is directly propotional to its circumference or radius as circular pitch or module of each gear is assured to be same.

$$\text{Speed ratio} = \frac{\text{Product of number of teeth on driven gears}}{\text{Product of number of teeth on drivers}}$$

or
$$\frac{N_1}{N_4} = \frac{T_2 \times T_4}{T_1 \times T_3}$$

Example 10.1. **A simple gear train has three gears each mounted on separate parallel shafts. Gear *A* is the driver and rotates clockwise at 750 rpm. It has *B* as intermediate gear and *C* is the follower. The number of teeth on gears *A, B* and *C* are 30, 45 and 75 respectively. Find**

(*a*) speed ratio of gear train.

(*b*) direction of rotation and speed of rotation of the follower.

Solution : The gear train is shown is the fig. 10.5.

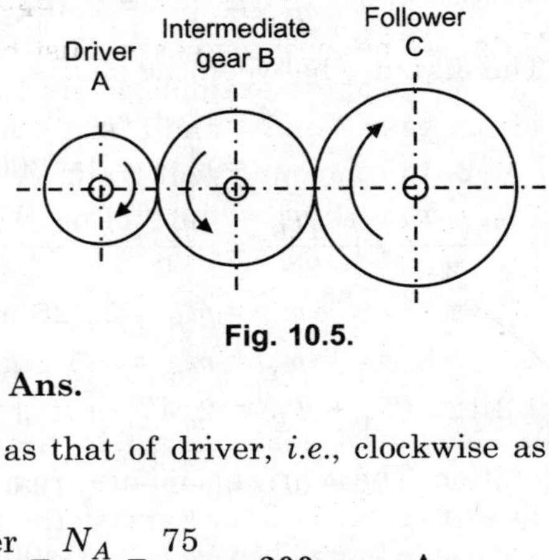

Fig. 10.5.

(*a*) Speed rotation

$$= \frac{\text{speed of driver}}{\text{speed of follow}} = \frac{T_C}{T_A} = \frac{75}{30} = \textbf{2.5 Ans.}$$

(*b*) The direction of follower is same as that of driver, *i.e.*, clockwise as shown

$$\text{Speed of follower} = \frac{\text{speed of driver}}{\text{speed ratio}} = \frac{N_A}{VR} = \frac{75}{2.5} = \textbf{300 rpm Ans.}$$

Example 10.2. **Power is transmitted from motor to output shaft through a compound gear train as shown. Find the direction and speed of output shaft if motor shaft rotate at 1250 rpm clockwise. The number of teeth are T_A = 30, T_B = 75, T_C = 20 and T_D = 50.**

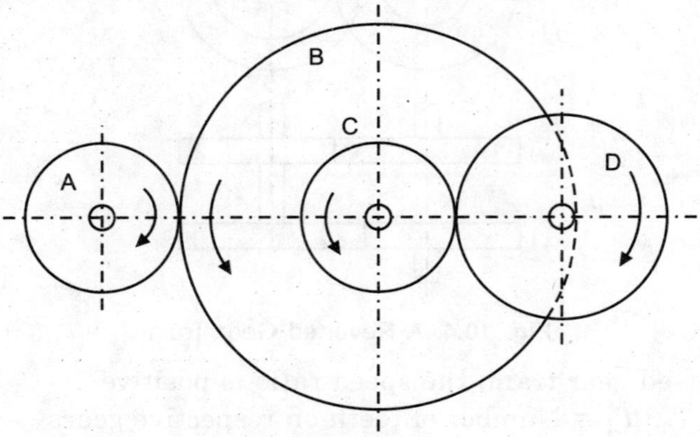

Fig. 10.6

Solution : $$\frac{N_A}{N_D} = \frac{T_B \times T_D}{T_A \times T_C} = \frac{75 \times 50}{31 \times 20} = 6.25$$

$$\therefore \quad N_D = \frac{N_A}{6.25} = \frac{1250}{6.25} = \textbf{200 rpm Ans.}$$

The direction of wheel D will be same as that of A, *i.e.*, clockwise.

Example 10.3. **A reverted gear train has a speed ratio of 12. The module pitch of gears A and B is 3.125 mm and that of gears C and D is 2.5 mm. The minimum number of gear teeth is 24. Calculate suitable number of teeth for the gears.**

Solution : $$\frac{N_A}{N_B} = \frac{N_C}{N_D} = \sqrt{12} = 3.464$$

$$\therefore \quad \frac{T_B}{T_A} = \frac{T_D}{T_C} = 3.464$$

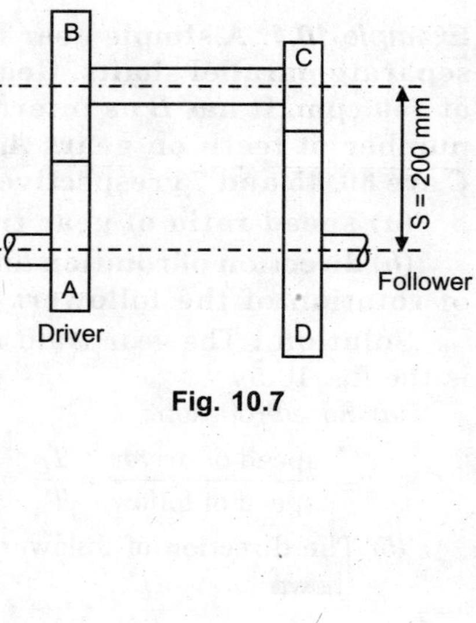

Fig. 10.7

The distance between the shafts

$$S = r_A + r_B$$

$$= r_C + r_D = 200 \text{ mm}$$

$$\therefore \frac{m_A \cdot T_A}{2} + \frac{m_B T_B}{2} = \frac{m_C T_C}{2} + \frac{m_D T_D}{2} = 200$$

$$m_A = m_B = 3.125 \text{ mm}$$

$$m_C = m_D = 2.5 \text{ mm}$$

$$\therefore 3.125\ (T_A + T_B) = 2.5(T_C + T_D) = 400$$

$$T_A + T_B = \frac{400}{3.125} = 128$$

$$T_C + T_D = \frac{400}{2.5} = 160$$

But $\quad T_B = 3.464\ T_A$

$\therefore \quad T_A + 3.464\ T_A = 128$

$$T_A = \frac{128}{(1+3.464)} = \frac{128}{4.464} = 28.67 \text{ say } 28$$

$$T_B = 128 - 28 = 100$$

Similarly, $\quad T_D = 3.464\ T_C$

$\therefore \quad T_C + 3.464\ T_C = 160$

$$T_C = \frac{160}{(1+3.464)} = \frac{160}{4.464} = 35.84 \text{ say } 36.$$

$$T_D = 160 - 36 = 124$$

The actual speed ratio,

$$\frac{N_A}{N_D} = \frac{T_B \times T_D}{T_A \times T_C} = \frac{100 \times 124}{28 \times 36} = 12.3.$$

10.6. PLANETARY OR EPICYCLIC GEAR TRAIN

An epicyclic gear train of two wheels is shown in Fig. 10.8. The axis of gear *A* is fixed and axis of gear *B* is free to rotate about the axis of gear *A* alongwith the arm. The arm holds the two gears together.

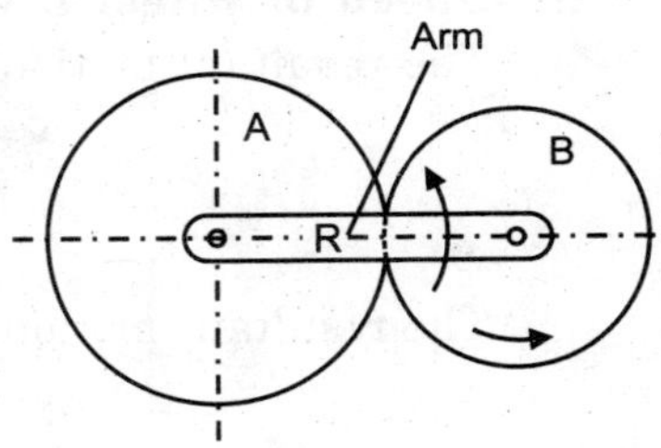

Fig. 10.8. Epicyclic Gear Train

If the arm *R* is fixed, the axes of both gears will be fixed automatically and this will become a simple gear train. The motion of various elements of gear train can be determined by tabular method as explained in Table 10.1.

Let T_A and T_B be be the number of teeth of gears *A* and *B*.

The values of *x* and *y* can be calculated depending upon the condition given in the problem. Hence motion of any element can be found out. The direction of motion of gear *B* is opposite to that of gear *A* and hence taken with –ve sign.

Table 10.1. Epicyclic Gear Train

Sl. No.	*Steps of motion*	*Revolution of Elements*		
		Arm R	*Wheel A*	*Wheel B*
1.	Arm is fixed. Rotate *A* through +1 revolution anticlockwise	0	+1	$-\frac{T_A}{T_B}$
2.	Arm is fixed. Rotate *A* through +*x* revolution anticlockwise	0	$+x$	$-x \cdot \frac{T_A}{T_B}$
3.	Add + *y* revolution to all elements	$+y$	$+y$	$+y$
4.	Resultant motion (Add second and third row)	$+y$	$x + y$	$-x\frac{T_A}{T_B} + y$

Example 10.4. **The arm of an epicyclic gear train carries two wheels *A* and *B* having 36 and 45 teeth respectively. The wheel *A* is fixed and the arm rotates about the centre of wheel *A* at 100 rpm anticlockwise. Find the speed of wheel *B*.**

If wheel *A* rotates at a speed of 200 rpm clockwise find the speed of wheel *B*.

Solution : Prepare the Table of Epicyclic Gear Train as per Table 10.1.

(*i*) Speed of wheel *B* when wheel *A* is fixed

Speed of arm R given = 100 rpm anticlockwise

$\therefore \quad y = +100$ rpm

Wheel A is fixed, $N_A = 0$

The resultant motion of wheel $A = x + y = 0$

$\therefore \quad x = -y = -100$ rpm

The resultant motion of wheel $B = -x\dfrac{T_A}{T_B} + y$

The speed of wheel B, $N_B = -x\dfrac{T_A}{T_B} + y = -(-100)\dfrac{36}{45} + 100$

$= 80 + 100 =$ **180 rpm** anticlockwise

(*ii*) Speed of wheel *B* when $N_A = -200$ rpm

The resultant motion of wheel A,

$$N_A = x + y = -200 \text{ rpm}$$

$$y = +100 \text{ rpm}$$

$$\therefore \quad x = -200 - y = -200 - 100 = -300$$

The resultant motion of wheel B

$$N_{Bs} = -x\frac{T_A}{T_B} + y = -(-300) \times \frac{36}{45} + 100$$

$$= 240 + 100 = \textbf{340 rpm anticlockwise}$$

Example 10.5. **In a reversed epicyclic train, the arm *R* carries two wheels *A* and *D* and a compound wheel *B* – *C* such that wheel *A* meshes with wheel *B* and wheel *D* mdshes with wheel *C*. The number of teeth are $T_A = 80$, $T_D = 48$ and $T_C = 72$. Calculate the speed and rotation direction of wheel *D* when wheel *A* is fixed and arm *R* rotates at 200 rpm clockwise.**

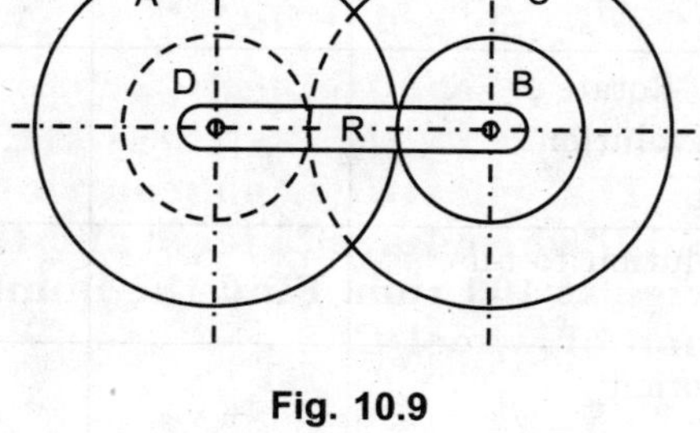

Fig. 10.9

Solution: Prepare the table of motion as follows:

Sl. No.	*Steps of motion*	*Speed of elements*			
		Arm R	*Wheel A*	*Compound Wheel B–C*	*Wheel D*
1.	Arm R is fixed. Rotate wheel A through +1 revolution anti-clockwise	0	+ 1	$-\frac{T_A}{T_B}$	$-\frac{T_A}{T_B}\times\left(-\frac{T_C}{T_D}\right)$ $=\frac{T_A}{T_B}\times\frac{T_c}{T_D}$
2.	Arm R is fixed Rotate wheel A through $+x$ revolution	0	$+ x$	$-x\frac{T_A}{T_B}$	$x\times\frac{T_A}{T_B}\times\frac{T_C}{T_D}$
3.	Add $+y$ revolutions to all elements	$+ y$	$+ y$	$+ y$	$+ y$
4.	Resultant motion	$+ y$	$x + y$	$-x\frac{T_A}{T_B}+y$	$x\frac{T_A}{T_B}\times\frac{T_C}{T_D}+y$

The resultant motion of arm R,

$y = -200$ rpm, (Arm rotats at 200 rm clockwise)

The resultant motion of wheel A

$x + y = 0$ (wheel A is fixed)

$\therefore \quad x = -y = -(-200) = 200$

The resultant motion of wheel D

$$= x\frac{T_A}{T_B}\times\frac{T_C}{T_D}+y$$

$$\therefore \quad N_D = x\frac{T_A}{T_B}\times\frac{T_C}{T_D}+y$$

$$= 200\times\frac{80}{40}\times\frac{72}{48}-200$$

$$= 600 - 200$$

= **400 rpm anticlockwise**

Example 10.6. An epicyclic gear train has a fixed annulus wheel *A* with 150 teeth. *A* wheel *B* meshes with *A* and drives a wheel *D* through an idler *C*. The wheel *D* is concentric with *A*. An arm *R* carriers wheels *B* and *D* which have 25 teeth and 40 teeth respectively and rotates clockwise at 100 rpm. Find the number of teeth, speed and sense of rotation of wheel *C*.

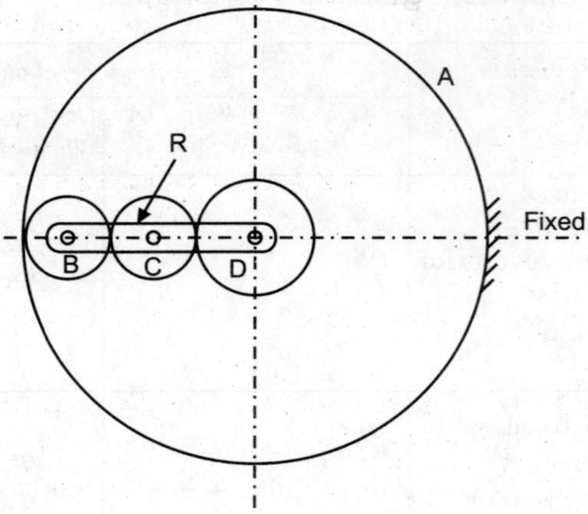

Fig. 10.10

Solution : Prepare table of motion

Sl. No.	*Steps of motion*	*Speed of elements*				
		Arm R	*Wheel A*	*Wheel B*	*Wheel C*	*Wheel D*
1.	Arm fixed, give one revolution to wheel *A* anti-clockwise	0	+1	$+\frac{T_A}{T_B}$	$-\frac{T_A}{T_B}\times\frac{T_B}{T_C}=\frac{T_A}{T_C}$	$\left(-\frac{T_A}{T_C}\right)\times\left(-\frac{T_C}{T_A}\right)=\frac{T_A}{T_D}$
2.	Multiply by x	0	$+x$	$+x\frac{T_A}{T_B}$	$-x\frac{T_A}{T_C}$	$+x\frac{T_A}{T_D}$
3.	Add y to all	$+y$	$x + y$	$+x\frac{T_A}{T_B}+y$	$-x\frac{T_A}{T_C}+y$	$+x\frac{T_A}{T_D}+y$

The resultant motion of arm R,

$y = -100$ [Arm rotates at 100 rpm clockwise]

The resultant motion of wheel A,

$x + y = 0$ [A is fixed]

$x = -y -(-100) = 100$

The resultant motion of wheel $C = -x\frac{T_A}{T_C} + y$

$$N_C = -x\frac{T_A}{T_C} + y$$

$$= -100\times\frac{150}{30}+(-100) \quad [\because\ T_A = 150,\ \ T_C = 30]$$

$$= -500 - 100 = -600 \text{ rpm}$$

= **600 rpm clockwise**

Example 10.7. **An epicyclic gear consists of a pinion, a wheel of 40 teeth and an annulus with 84 internal teeth concentric with the wheel. The pinion gears with the wheel and the annulus. The arm that carries the axis of the pinion rotates at 100 rpm. If the annuls is fixed, find the speed of the wheel. If the wheel is fixed, find the speed of annulus.**

[UPTU 2010-11]

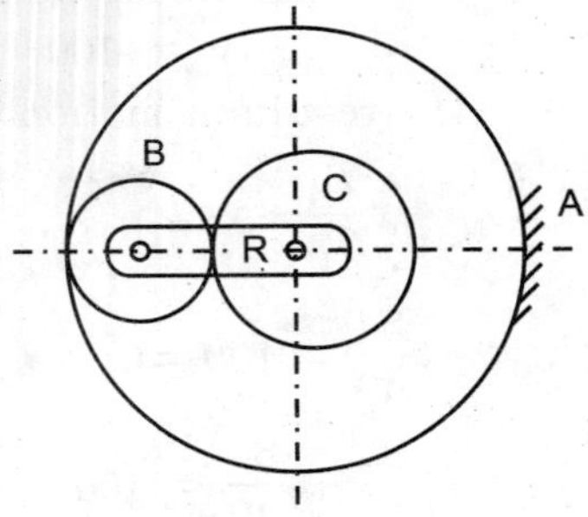

Fig. 10.11

Solution :

$$T_C = 40$$
$$T_A = 84$$
$$N_R = 100 \text{ rpm}$$

Prepare the table of motion

Sl.No.	*Steps of motion*	*Speed of elements*			
		Arm R	*Wheel A*	*Wheel B*	*Wheel C*
1.	Arm fixed give one revolution to wheel A anti-clockwise	0	+ 1	$-\frac{T_A}{T_B}$	$-\frac{T_A}{T_B} \times \frac{T_B}{T_C} = -\frac{T_A}{T_C}$
2.	Multiply by x	0	$+ x$	$+x\frac{T_A}{T_B}$	$-x\frac{T_A}{T_C}$
3.	Add $+ y$ to all	$+ y$	$x + y$	$+x\frac{T_A}{T_B} + y$	$-x\frac{T_A}{T_C} + y$

(*a*) The resultant motion of arm R

$y = -100$ (Arm rotates at 100 rpm clockwise assumed)

The resultant motion of wheel A,

$$x + y = 0$$
$$x = -y = -(-100) = 100$$

The resultant motion of pinion $B = x\frac{T_A}{T_B} + y$

$$\therefore \quad N_B = x\frac{T_A}{T_B} + y$$
$$= 100 \times \frac{84}{T_B} - 100 = \frac{84}{T_B}$$

The resultant motion of wheel $C = -x\frac{T_A}{T_C} + y$

The speed of wheel C

$$N_C = -x\frac{T_A}{T_C} + y = -100 \times \frac{84}{40} - 100$$
$$= -210 - 100 = -310 = 310 \text{ rpm clockwise}$$

(*b*) The resultant motion of arm *R*,

$$y = -100$$

The resultant motion of wheel *C*,

$$-x\frac{T_A}{T_C} + y = 0$$

$$-x\frac{84}{40} - 100 = 0$$

$$-x\frac{84}{40} = 100$$

$$\therefore \quad x = -\frac{100 \times 40}{84} = -47.62$$

The resultant motion of wheel $A = x + y$

$$N_A = x + y = -47.62 - 100 = -147.62$$

$$= \mathbf{147.62\ rpm\ clockwise}$$

10.7. SUN AND PLANET GEAR

It is a modified epicyclic gear train with the following features.

1. The gear *A* co-axial with the annulus gear *C* is called the sun gear. This gear can be fixed or can rotate freely.
2. The planet gear *B* can revolve round the sun gear *A*. It meshes with annular gear *C*. It is carried by arm *R* and can rotate freely.
3. The annular gear *C* has interval teeth. It can be fixed or can rotate freely.
4. **Mode of operation**
   The aim *R* is always the driven member. If gear *A* is fixed, the gear *C* is the driving member. If gear *C* is fixed, the gear *A* is the driving member.
5. **Types of sun and planet gears.**
   (*a*) Simple train

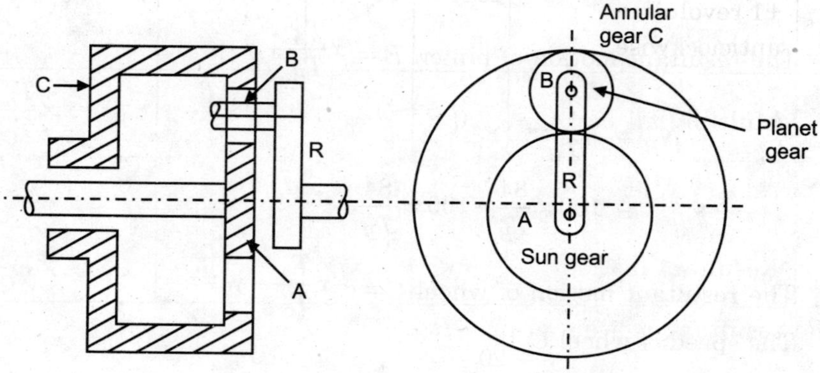

Fig. 10.12. Sun and Planet Simple Train

(*b*) Compound train

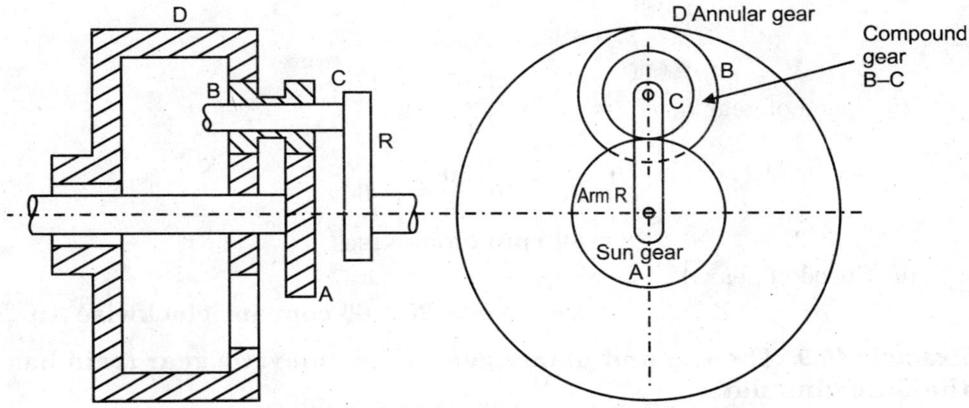

Fig. 10.13. Sun and Planet compound Train

6. The method of solution (design) is similar to that used for epicyclic gear train.

Example 10.8. **The sun and planet gear of an epicyclic gear train has the following data.**

No. of teeth of annular gear, T_C = 100 (internal)

No. of teeth of sun gear, T_A = 50 (external)

No. of teeth of pinion, T_B = 25 (external)

The gear *C* is fixed and arm *R* rotates at 20 rpm. Find the speed of gears *A* and B.

Solution : Prepare the table of motion

Sl.No.	*Operation*	*Speed of elements*			
		Arm R	*Wheel A*	*Wheel B*	*Wheel C*
1.	Arm fixed, rotate sun gear *A* though +1 revolution anticlockwise	0	+1	$-\frac{T_A}{T_B}$	$-\frac{T_A}{T_B} \times \frac{T_B}{T_C} = -\frac{T_A}{T_C}$
2.	Multiply all by x	0	$+x$	$-x \times \frac{T_A}{T_B}$	$-x \times \frac{T_A}{T_C}$
3.	Add $+y$ to all	$+y$	$x + y$	$+y$	$+y$
4.	Resultant motion	$+y$	$x + y$	$y - x\frac{T_A}{T_B}$	$y - x\frac{T_A}{T_C}$

The resultant motion of airm *R*

$$y = 20$$

The resultant motion of annular gear *C*,

$$y - x \times \frac{T_A}{T_C} = 0 \qquad \text{[wheel } C \text{ is fixed]}$$

$$y - x \times \frac{50}{100} = 0$$

$$x = 40$$

(*i*) Speed of gear *B*, $N_B = y - x\ \frac{T_A}{T_B}$

$$= 20 - 40\ \frac{50}{25} = -60$$

$$= \mathbf{60\ rpm\ clockwise}$$

(*ii*) Speed of gear *A*

$N_A = x + y = 40 + 20 =$ **60 rpm anticlockwise Ans.**

Example 10.9. **The sun and planet gear of an epicyclic gear train has the following data.**

The number of teeth of sun gear, $T_A = 60$

The number of teeth of driven gear of compound gear $T_B = 40$

The number of teeth of driving gear of compound gear, $T_C = 25$

The speed of shaft connected to gear *A*, $N_A = 100$ rpm

The annular gear *D* is fixed and all gears have same module.

Find the number of teeth of annular gear *D* and speed of the shaft connected to the arm.

Solution : Prepare the table of motion

Sl.No.	*Operation*	*Speed of elements*			
		Arm R	*Gear A*	*Compound gear B-C*	*Gear D*
1.	Arm fixed, give gear *A* a rotation of +1 anticlockwise	0	+1	$-\frac{T_A}{T_C}$	$-\frac{T_A}{T_C} \times \frac{T_B}{T_D}$
2.	Multiply all by +*x*	0	+*x*	$-x \times \frac{T_A}{T_C}$	$-x \times \frac{T_A}{T_C} \times \frac{T_B}{T_D}$
3.	Add +*y* to all	+*y*	+*y*	+*y*	+*y*
4.	Resultant motion	+*y*	*x* + *y*	$y - x\frac{T_A}{T_C}$	$y - x\frac{T_A}{T_C} \times \frac{T_B}{T_D}$

The resultant speed of gear $D = y - x\frac{T_A}{T_C} \times \frac{T_B}{T_D}$

$$\therefore \quad N_D = y - x\frac{T_A}{T_C} \times \frac{T_B}{T_D} = 0$$

$$y - x\frac{60}{25} \times \frac{40}{125} = 0$$

$$\therefore \quad y - 0.768x = 0$$

The resultant speed of gear $A = x + y$

$$N_A = x + y = 100$$

$$x = 100 - y$$

$$\therefore \quad y - 0.768\,(100 - y) = 0$$

$$y = 43.43$$

$$\therefore \quad N_R = \mathbf{43.43\ rpm\ anticlockwise}$$ **Ans.**

10.8. TORQUES AND TOOTH LOADS IN EPICYCLIC GEAR TRAINS

All gears in epicyclic gear train move with uniform speeds and therefore, have no angular acceleration. The gear train is in equilibrium under the action of following three external torques.

(*i*) Driving or input torque (T_1)

(*ii*) Output torque (T_2) of the driving element

(*iii*) Holding or braking or fixing torque (T_3)

For equilibrium,

$$T_1 + T_2 + T_3 = 0$$

or $$F_1 \times r_1 + F_2 \times r_2 + F_3 \times r_3 = 0$$

where F_1, F_2, F_3 are external forces and r_1, r_2 and r_3 are corresponding radii of gears.

Assuming no losses in power transmission due to friction, the net energy dissipated by the gear train will be zero

$$T_1\omega_1 + T_2\omega_2 + T_3\omega_3 = 0$$

where ω_1, ω_2, ω_3 are corresponding angular velocities of the train elements.

For fixed or holding element, $\omega_3 = 0$

$$\therefore \quad T_1\omega_1 + T_2\omega_2 = 0$$

$$T_2 = -\frac{T_1\omega_1}{\omega_2}$$

$$\therefore \quad T_1 + \left(-\frac{T_1\omega_1}{\omega_2}\right) + T_3 = 0$$

or $$T_3 = -\ T_1 + \frac{T_1\omega_1}{\omega_2} = T_1\left(\frac{\omega_1}{\omega_2} - 1\right)$$

If input torque T_1 is known, the holding torque T_3 can be estimated.

QUESTION BANK

1. What are the differences between a simple gear train and reverted gear train? Explain with the help of sketches.
2. What are main types of gear trains ? Explain with the help of suitable sketches.
3. Explain the procedure for the analysis of an epicyclic geartrain.
4. What is a sun and planet gear? How are these gears analysed?
5. What is braking or fixing torque of a gear in an epicyclic gear train? How is it estimated?

NUMERICAL PROBLEMS

1. In a sun and planet gear train, the sun gear wheel having 60 teeth is fixed. The annulus wheel and arm rotate at 130 rpm and 100 rpm in the same direction. Determine the number of teeth on the planet and annular wheels.

   [Ans: 70, 200]

2. An epicyclic gear train consists of an annulus of 84 teeth, a wheel of 40 teeth and a pinion meshes with both annulus and wheel. The arm carrying the pinion rotates at 100 rpm. If the annulus is fixed, find the speed of the wheel. If wheel is fixed, find the speed of the annulus. **[Ans:** 310 rmp, 147.6 rpm]

3. Two spur gears of an epicyclic gear train have 24 and 30 teeth respectively. The arm rotates at 100 rpm in clockwise direction. If one gear is fixed find the speed of other gear. **[Ans:** 180 rpm clockwise]

❑❑❑

Annexure—I

UNIVERSITY PAPER

B.Tech.

(SEM. V) ODD SEMESTER THEORY EXAMINATION, 2011-12

THEORY OF MACHINES-I

Time: 3 Hours *Total Marks: 100*

Note: 1. There are total five questions. Attempt all questions.
2. Marks are indicated against each question.
3. Assume missing data suitably, if any.

1. Attempt any four of the following: — **(4 × 5 = 20)**

(*a*) Define the following:
Link, resistant body, kinematic pair, mechanism and inversion of mechanism.

(*b*) Describe the function of crank and slotted lever mechanism.

(*c*) State and prove the Kennedy's theorem as applicable to instantaneous centres of rotations of three bodies.

(*d*) Explain the procedures to construct Klein's construction to determine the velocity and acceleration of a slider crank mechanism.

(*e*) Show that a pantograph can produce paths exactly similar to the ones traced out by a point on a link on an enlarged and reduced scale.

(*f*) The ratio between the width of the front axle and that of the wheel base of a steering mechanism is 0.44. At the instant when the front inner wheel is turned by 18°, what should be the angle turned by the outer front wheel for perfect steering?

2. Attempt any two of the following :

(*a*) Show that an Ackermann steering gear does not satisfy the fundamental equation of steering gear at all positions. Why is it widely used ? (10)

(*b*) (*i*) What is velocity of rubbing ? How is it found ? (3)

(*ii*) For the position of the mechanism shown in Fig. 1, calculate the angular velocity of link AR. OA is 300 mm and rotates at 20 rad/s in the clockwise direction. (7)

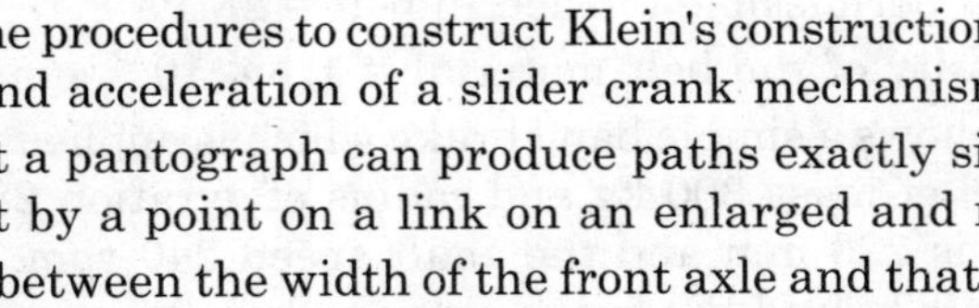

Fig. 1

(c) For the inverted slider crank mechanism shown in Fig. 2, find the angular acceleration of link QR. The crank OA is 300 mm long and rotates at 20 rad/s in clockwise direction. (10)

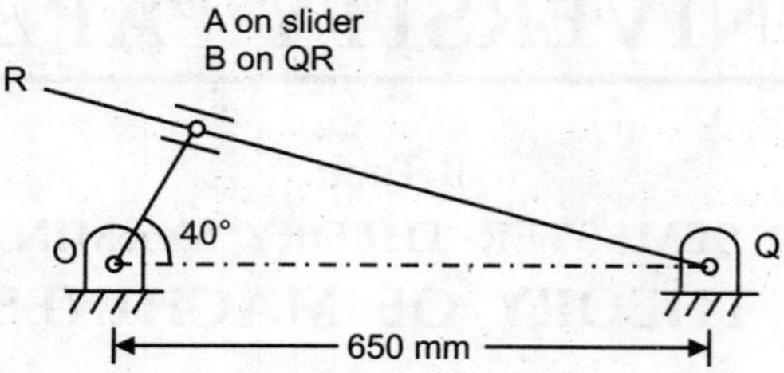

Fig. 2

3. **Attempt any two of the following :**

(a) A thrust bearing of a propeller shaft consists of a number of collars. The shaft is of 400 mm diameter and rotates at a speed of 90 rpm. The thrust on the shaft is 300 kN. If the intensity of the pressure is to be 200 kN/m^2 and coefficient of friction is 0.06, determine external diameter of the collars. The power lost in friction is not to exceed 48 kW. (10)

(b) Determine the maximum power that can be transmitted through a flat belt having the following data :
Cross-section of the belt is = 300 mm × 12 mm
Ratio of friction tensions = 2.4
Maximum permissible tension in belt = 2MPa
Mass density of the belt material = 1.1×10^{-3} g/mm^3. (10)

(c) Figure 3 shows a simple band brake which is applied to a shaft carrying a flywheel of mass 300 kg and radius of gyration 280 mm. The drum diameter is 220 mm and the shaft speed 240 rpm. The coefficient of friction is 0.3. Find the brake torque when a force of 100 N is applied at the lower end. Also determine the number of turns of the flywheel and time taken by it before coming to rest. All dimensions are in mm. (10)

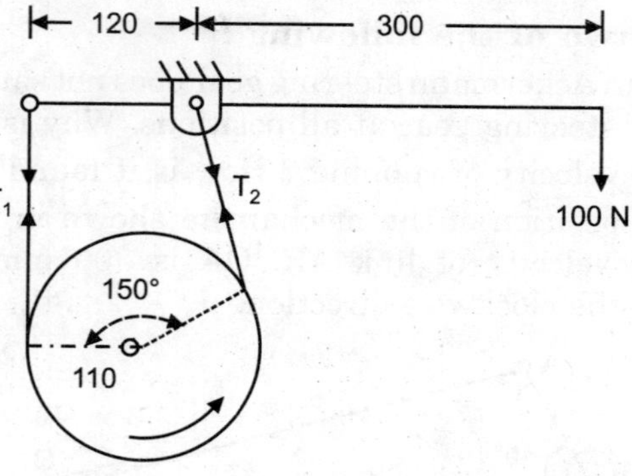

Fig. 3

4. **Attempt any two of the following :**

(a) (i) State and derive the law of gearing. (7)

(ii) What are the differences between a simple gear train and reverted gear train ? Explain with the help of sketches. (3)

(b) A pinion of 20° involute teeth rotating at 275 rpm meshes with a gear and provides a gear ratio of 1.8. The number of teeth on the pinion is 20 and the module is 8 mm. If the interference is just avoided, determine:

(i) the addenda on the wheel and the pinion

(ii) the path of contact, and

(iii) the maximum velocity of sliding on both sides of the pitch point. (10)

(c) An epicyclic gear consists of a pinion, a wheel of 40 teeth and an annulus with 84 internal teeth concentric with the wheel. The pinion gears with the wheel and the annulus. The arm that carries the axis of the pinion rotates at 100 rpm. If the annulus is fixed, find the speed of the wheel; if the wheel is fixed, find the speed of the annulus. (10)

5. Attempts any two of the following :

(a) (i) Deduce the expression for the velocity and acceleration of the follower when it moves with simple harmonic motion. Why simple harmonic motion program is not suitable for high speed cams(7)

(ii) What do you understand by undercutting of the cam? (3)

(b) Use the following data in drawing the profile of a cam in which a knife-edged follower is raised with uniform acceleration and deceleration and is lowered with simple harmonic motion :

Least radius of cam = 60 mm.

Lift of follower = 45 mm

Angle of ascent = 60°

Angle of dwell between ascent and descent = 40°

Angle of descent = 75°

If the cam rotates at 180 rpm, determine the maximum velocity and acceleration during ascent and descent. (10)

(c) A tangent cam with straight working faces tangential to a base circle of 120 mm diameter has a roller diameter of 48 mm diameter. The line of stroke of the roller follower passes through the axis of the cam. The nose circle radius of the cam is 12 mm and the angle between the tangential faces of the cam is 90°. If the speed of the cam is 180 rpm, determine the acceleration of the follower when :

(i) During the lift, the roller just leaves the straight flank.

(ii) The roller is at the outer end of its lift, *i.e.*, at the top of the nose. (10)

□□□

Annexure—II

REFERENCE BOOKS

1. **Theory of Machines Thomas Bevan,** *CBS Publishers and Distributors, Delhi.*
2. **Theory of Machines and Mechanisms- Shigley,** *Uickers, Pennock, Oxford University Press, N. Delhi*
3. **Theory of Machines and Mechanisms**-*Ghosh and Mallik, Affliated East-Wat Press. Pvt. Ltd. New Delhi*
4. **Theory of Machines and Mechanisms** - *Rao and Dukkipati*
5. **Theory of Machines-** *S.S. Rattan, Tara McGraw Hill Publishing Co. Ltd., New Delhi*
6. **Kinematics of Machines**-*Dr. Sadhu Singh*
7. **Mechanics of Machines**-*V.Ramamurti*
8. **Theory of Machines** - *Khurmi and Gupta, Eurasia Publishing House Pvt. Ltd. New Delhi*
9. **Theory of Machines-** *R.K. Bansal, Lakshmi Publications Pvt. Ltd., New Delhi.*
10. **Theory of Machines**-*V.P. Singh*
11. **Theory of Machines**-*Malhotra and Gupta.*

Index

Index...

K...

L...

M...

O...

P...

R...

S...

T...

U...

V...

W...

□□□

Win Prizes

ATTENTION READERS

Win Prizes

We request for your frank assessment regarding various aspects of the book as given below :

THEORY OF MACHINES-I

By: R.K. SINGAL

Please Fill up the Sheet in Neat and Clean Writing

(i) What topics of your syllabus, important for your examination point of view, are not covered in this book ?

....

....

....

....

(ii) What are the chapters wherein the treatment of the subject matter is not systematic or organised or updated ?

....

....

....

....

(iii) Have you come across misprints/mistakes/factual inaccuracies in the book? Please specify the chapters and the page numbers.

....

....

....

....

(iv) What other books on the same subject have you found/heard better than the present book? Please specify in terms of price and quality.

....

....

....

....

(v) Further suggestions and comments for the improvement of this book.

....

....

....

....

(PS: If need be please attach a separate sheet to write your views)

OTHER DETAILS

(i) Who recommended you this book ? (Please tick (✓) in the box near the option relevant to you.

☐ Teacher ☐ Friends ☐ Bookseller

....

....

....

....

(ii) Name of the recommending teacher, his designation and address :

....

....

....

....

(iii) Name and address of the bookseller you purchased the book from :

....

....

....

....

(iv) Name and address of your institution (Please mention the University or Board as the case may be)

....

....

....

....

(v) Your name and complete postal address :

....

....

....

....

(vi) Course you are enrolled in :

....

....

....

....

The best assessment will be awarded each month: The award will be in the form of our publications as decided by the Editorial Board.

Please mail the filled up coupon at your earliest to :

S.K. KATARIA & SONS®
4760-61/23, Ansari Road,
Daryaganj, New Delhi-110002
Phone: +91-11-23243489, 23269324
Telefax: +91-11-23243489
e-mail : katariabooks@yahoo.com
katariabook@gmail.com